Urbanisation, Development and Environment

Urbanisation, Development and Environment

C.D. Deshpande Memorial Volume

Editors
V.S. PHADKE
S. BANERJEE-GUHA

RAWAT PUBLICATIONS
Jaipur • New Delhi • Bangalore • Mumbai • Hyderabad • Guwahati

ISBN 81-316-0056-4

Published by
Prem Rawat for **Rawat Publications**
Satyam Apts., Sector 3, Jawahar Nagar, Jaipur 302 004 (India)
Phone: 0141 265 1748 / 7006 Fax: 0141 265 1748
E-mail: info@rawatbooks.com
Website: www.rawatbooks.com

New Delhi Office
4858/24, Ansari Road, Daryaganj, New Delhi 110 002
Phone: 011-23263290

Also at Bangalore, Mumbai, Hyderabad and Guwahati

Typeset by Rawat Computers, Jaipur
Printed at Chaman Enterprises, New Delhi

Contents

Professor C.D. Deshpande

A Profile

Professor C.D. Deshpande, born on 14th January 1912, studied Economics, Politics and History as his basic subjects for his degree in the University of Mumbai. Subsequently, he did his diploma in Geography from the University of California. On his return, he wrote a book titled *Western India* for which he was awarded the D. Lit. degree of the University of Mumbai.

Prof. Deshpande's professional career started in Karnataka College, Dharwar, as a lecturer in Geography, and he later became the principal there. Being in the service of the government of Bombay Province, he was subsequently transferred to Visnagar College, Visnagar, Gujarat. He was a born teacher and this was reflected in the clarity of his thoughts while teaching. A person of even average intelligence could follow what he taught without much difficulty, irrespective of whether he was treading on the giddy heights of theories in geomorphology, or walking in the plains of the man-environment relationship, or flowing in the fields of Geographical Thought or Regional Geography. Prof. Deshpande was a teacher with great insight and all-embracing vision. His teaching was simple, lucid and straightforward. His audiences always felt dazed by the way he developed a concept and systematically built up a theme, using simple expressions, strictly avoiding jargon, and sprinkling humour to make his teaching comfortable. Nobody used to get tired or bored at the end of his lectures. His lectures were gently flowing rivers nourishing the students at every point. He fitted into any crowd he came across.

From teaching, Prof. Deshpande shifted to administration and became the chairman of the Maharashtra State S.S.C. Board. In recognition of his useful service as an able administrator, he was appointed the additional director and finally the director of Education, Maharashtra state. Although he was busy streamlining educational administration in the newly reorganised state of Maharashtra, he never lost his love for teaching and research in Geography. He used to frequently visit the post-graduate departments, especially in Parle College, now Sathye College, where he taught whenever the situation demanded. He could keep his research interests alive despite his busy schedule.

Prof. Deshpande retired from the post of director of Education in November 1969 and immediately joined the University of Mumbai as the professor and founder head of its Department of Geography, a post which he held for seven years until 1976. While holding this post, he had to face several problems, such as shortage of faculty, non-teaching staff and space, which got accentuated when the department was shifted to the Vidyanagari campus, a physical as well as social wilderness at that time. He accepted all these challenges and laid down a solid foundation for the discipline. The department grew from strength to strength and the credit for its initial expansion goes solely to him.

After retirement, Prof. Deshpande was awarded a senior fellowship of the ICSSR for three years. During this period, he was attached to the Jawaharlal Nehru University and authored a book titled *India: A Regional Reinterpretation.* While in Delhi, he took the initiative to establish the National Association of Geographers, India, (NAGI) in 1979, which finally became the leading professional organisation of geographers at the national level. He was the founder president of this association, and, later, made the chief honorary editor of the *Annals*, the research journal of the association, for 10 years from 1981-91. Due to his sustained research interests, he was awarded the Fellowship of the Royal Geographical Society, London. He was also associated with the activities of the International Geographical Union for a long time.

Prof. Deshpande had an illustrious professional career. He was associated with a number of professional bodies such as the Indian Geographical Society, Deccan Geographical Society and Bombay Geographical Association. He was a member of the Marathi Viswakosh Mandal and the Board of University Textbook Production, Nagpur. He wrote a book titled *Shahare* in Marathi for this board, which can be considered a textbook in Urban Geography at the post-graduate level. The book, titled *Geography of Maharashtra*, which he wrote for the National Book Trust during his tenure in the University of Mumbai, is a widely referred book by both students and teachers. He was also invited to work as a visiting faculty in the Military Engineering College, Pune, and the Institute of Town Planning, Pune, where he taught Regional Planning for several years. He contributed to several district gazetteers of Maharashtra. Even in the last days of his life, he was busy authoring several articles for *Encyclopaedia Britannica.* Despite all these assignments, however, he kept in close contact with primary education by serving on the academic committee related to the production of geography textbooks of the Maharashtra State Bureau of Textbook Production and Curriculum Research, Pune.

Prof. C.D. Deshpande was a man with a humanistic approach, who strived to understand the complex situations of human life. In spite of his personal sorrows, he never lost interest in life and remained very active till the end. He was always cheerful and bubbling with energy and enthusiasm. He was a voracious reader, a man with vision, and a remarkable person. An erudite Sanskrit scholar, Prof. Deshpande was a multifaceted personality, always inspiring children, youth and elders alike. With big ears and a bigger heart, he was ever accommodating, forgiving and tolerant. At the same time, he was as deep as the ocean, vast as the sky, focused and working for a large part of the day till the end of his life.

He passed away on 14th November 1999 in Pune. We revere the 'teacher' in him, who unraveled the beautiful vision of the unified perspective of geography to many of us.

V.S. Phadke
S. Banerjee-Guha

Introduction

Historically, urbanisation has been viewed as an important factor of economic change, inducing the breakdown of the feudal order and making society move towards a higher social formation. In the early phases of industrial growth and expansion, urbanisation did fulfil three major economic roles by (1) inducing labour shift from agrarian areas to the expanding urban industrial areas; (2) absorbing surplus labour from the agricultural sector and increasing labour productivity; and (3) diffusing technological innovation from the above centres across space, creating conditions for the development of non-industrial sectors. Naturally, cities associated with such processes were generators of economic growth and known as generative cities.

The above relationship between urbanisation and development was not found to be replicated in the case of Third World countries and no analogy actually can be drawn between these two sets of countries. Factors of transition and convergence did not give rise to similar patterns in industrial and less developed countries because of the differential roles of capitalist development in these two sets of countries, which got intrinsically associated with the expansion of urbanisation. Due to the differentials in resource and population distribution in the policies followed by the governments, and also due to the diverse historical relationships with international capital, the cities of the less developed countries could not fulfil similar roles as done by their counterparts in the Western world in the 19th century. As a result, the development pattern in space in these countries became uneven as also the urbanisation pattern, evident in the differentials among core cities, smaller urban centres and peripheral areas and between urban and rural areas. The rapid rate of urban growth or voluminous rural-urban migration expressed only a false analogy of form, but the process remained different. Inside these cities, poverty and deprivation became the common elements of spatial formations.

The inexorable process of urbanisation in these countries has become extremely dominant, significantly changing the parameters of development by impacting social, economic and environmental sustainability. It is estimated that the number of people living in urban areas all over the world would touch the five billion mark by 2025. The phenomenon is intrinsically linked and irrevocably intertwined with the development process, simultaneously creating inequalities of various kinds within populations and regions. The divide between centres and the surrounding rural areas become sharper with advancement in economy and, thereby, city growth. Urban areas are often equated with high development and rural, with low development.

As a natural corollary, large cities in these countries, due to their rapid population growth and concentration of economic activities, have become more a foci of environmental pollution and extracting points of regional resources rather than promoters of diffusion of growth. They essentially perform a parasitic role in the process of regional development.

Metropolitanisation is encouraged simultaneously by public policies as well as external forces, leading to extreme primacy in the settlement system. Degradation of the environment surfaces as a concomitant pattern of such urban growth, both inside the cities and in the surrounding rural areas.

In the light of the above, this book, containing eighteen articles, unfolds the complex process and pattern of urbanisation, development and the environmental situation, primarily in India, in longitudinal terms.

The articles in this book, dealing with a wide range of topics under urbanisation, development and environment at various spatial scales, and with a spatial perspective, have been broadly classified into seven groups. The first group of three articles delves into issues, general and particular, related to environment and sustainable development; the second group – a single article – deals with the problems of regional disparity in Russia in the context of the transition; the third group of two articles deals with economic and health issues at the national level; the fourth group article discusses economic and urbanisation issues related to Maharashtra; the fifth group of six articles focuses on administrative, environmental, historical and infrastructural aspects of states other than Maharashtra; the sixth, with two articles, deals with regional problems in Maharashtra, environmental and socioeconomic in nature; and the seventh group, with three articles, focuses on urbanisation and the economic issues of the city of Mumbai and its metropolitan region.

The three articles that have been included in the first group deal with issues of environment and sustainable development. In the first article, **'Environment, Sustainable Development and Tradition'**, **S.P. Das Gupta** initiates the discussion with a cautionary word that no economic development effort can be sustained under the current system of massive destruction and exploitation of scarce natural resources. According to Das Gupta, for any new developmental alternative to be the most suitable option for an area, should be formulated according to already established traditions. This, he opines, would not only conserve and protect the available environmental resources, but also enhance them. Much of the knowledge to support such a development alternative, Das Gupta suggests, lies in the traditional Indian approach towards the sustainable growth of the economy and material culture. However, he says that not all traditional methods are worthy enough; many have worked against nurturing biodiversity, especially those with the sole purpose of accumulation of wealth. The most viable answer, according to Das Gupta, lies in the promotion of ecofriendly traditional skills in different sectors of the economy.

In the same vein, but with a much wider perspective, **Parbati K. Sircar**, in his paper titled **'Managing Environmental Crisis'**, discusses global environmental problems in general and those faced by India in particular. According to Sircar, the major problems are global warming, depletion of the ozone layer, acid rain, etc., that have led to a global environmental crisis and, hence, warrant global environmental actions. It is true that over the last few decades, there has been an increasing awareness of global environmental problems and wide-ranging efforts to protect and conserve the environment. Sircar, in his paper, talks of significant endeavours such as the World Commission on Environment and Development, U.N. Conference on Environment and Development, U.N. Convention on Biodiversity, 1992, etc. He also classifies the major environmental problems in India under three categories related to (a) poverty and maldistribution of resources; (b) overpopulation, and (c) rapid industrialisation and urbanisation. He lists the significant measures taken by the legal system, more specifically the Supreme Court, and the people at large. He pins high hopes on people's efforts represented by NGOs, women's groups, tribal committees and other similar fora.

In the third paper of the group, titled **'A Systems Approach in Coastal Zone Management'**, **Satyesh C.**

Chakraborty takes the discussion on understanding environmental problems from the general to the particular level through an exposition of the management of environment in coastal zones. Enlightening the readers on the specific nature of the coastal zone, Chakraborty initially emphasises the task of managing such an environment and the steps to achieve this management. He recommends various pragmatic steps for altering the state of the environment in stages. With a detailed description of the nature of the coastal environment and its processes, Chakraborty unfolds the intricacies of the various regulation zones in the coastal ecosystem. Finally, he emphasises the need to develop a comprehensive understanding of such zones that are the most sensitive to human impact.

The solitary paper in the second group, titled **'Russia in Transition: Widening Regional Income Disparities'** by **R.G. Gidadhubli**, is of an entirely different nature as compared to the other papers in this volume. Being a researcher on issues of the economic development of the erstwhile Soviet Union and with his continuing interest in developmental patterns in the post-Soviet period, Gidadhubli has focused attention on the problems faced by Russia in its period of transition. He has vividly brought out the issues related to development of contemporary Russia in the context of its historical development during the Soviet era. The paper amply demonstrates how well-intentioned planning concepts such as regional specialisation, balanced and complex regional development, etc., backfired after the disintegration of the Soviet Union; even during the Soviet period, the success was at best partial because the regional development was subservient to national interests. Gidadhubli drives home the fact that disparities exist even in the states that are progressive constituents of the Commonwealth of Independent States (CIS) the post-Soviet period, and Russia is not an exception to this. The author lists certain crucial factors responsible for this state, including the changes in policy instruments by the contemporary regime, dismemberment of old links and the consequent drop in demand for certain commodities and the economic obsolescence of defence production units after the end of the Cold War. Gidadhubli finally draws attention to the social issues arising out of these economic disparities and the migration to large cities from depressed and crisis-ridden regions. Gidadhubli argues that although there are certain programmes to relieve these socio-economic problems, they have not served much beyond paying mere lip service.

The first article of the third group, '**Revisiting the Growth of Foreign Collaborations in Indian Industries in the Pre-liberalisation Era: A Spatial Analysis of 1970-90**', authored by **Swapna Banerjee-Guha**, is a critical analysis of the pattern of spatial growth of foreign collaborations in India during the pre-globalisation period of 1970-90. Following the debate on time-space convergence, Banerjee-Guha traces the historically oriented regional pattern of corporate locational dynamics initiated in India by a select group of parent countries during different time periods, according to the needs and claims of foreign capital. She identifies the 1980s as extremely significant with increasing foreign collaborations in metropolitan locations. This was also the period when Mumbai emerged as the biggest metropolitan drawer – an identity it had been developing since the late 1960s – of collaborative ventures in the country, a position that was challenged by Delhi only in later years. In this paper Banerjee-Guha deftly analyses the geographical seesaw of the decline and growth of the four regions and the respective metropolises in the country as locations of foreign collaborations and, thereby, lends a historical materialist approach to the analysis of the space economy of foreign collaborations. The phenomenal geographical concentration of foreign collaboration units is finally seen by her as the result of a heterogeneous combination of various factors of resource, labour, communication, investment incentives, political situation and specific social relations. The complex mix of these factors, on which she focuses in her analysis, brings out the significance of the political economic perspective developed since the late 1980s in Human Geography.

Using the applied interventionist approach, **Jayasree De**, in her paper, '**Urban Environment, Sanitation and Health in India**', critically analyses the role of the urban environmental conditions in India in contributing towards health hazards. Identifying Indian cities as centres of inmigration and, hence, congestion and crowding, De locates the sources of various contaminable diseases in the associated unhygienic conditions. With a brief description of the urban environment and sanitation in various size classes of cities, she elaborately discusses the root cause of various types of urban diseases. While tracing the cause-effect relationship of the diseases, she succinctly classifies them according to season, type of city and even specific urban land use, living conditions and lifestyle. Hence, intra-urban and inter-urban disparities emerge as significant issues in her analysis of urban health conditions in India. De opines that human intervention in the ecosystem has been largely responsible for the multiple nature of the health problems in Indian cities and sharply points at the degradation of the urban environment as caused even by family planning and ill-conceived policies. She suggests a multi-faceted approach for the restoration of urban health that can only be achieved through an integration of economic and social welfare schemes having a broad-based perspective towards urban infrastructural development.

The sole article of the fourth group, '**Urbanisation and its Socioeconomic Correlates in Maharashtra, 1991**', by **Dipti Mukherji and V.S. Phadke**, attempts to understand urbanisation in relation to the other socioeconomic dimensions of development in the state of Maharashtra for the year 1991. This has been done with the help of 21 key indicators employing the factor analytical technique to understand the spatial association of the variables. The analysis reveals that urbanisation is spatially related only to workers in some important non-agricultural activities like factory manufacturing, construction and various tertiary activities. Moreover, the level of urbanisation is primarily a contribution of large urban settlements. Medium-small town polarity stands as a separate orthogonal factor indicating the negligible role of these towns in the urbanisation process. They not only occupy different spatial segments when compared to cities, they also do not co-exist in themselves. Infrastructure and literacy do not show any strong relation to the urbanisation factor, but occupy different spatial segments on their own. Mukherji and Phadke contend that there is a strong influence of policy and administrative factors in the localisation of social overheads, apart from the influence of culture and tradition. Transport, which is also affected by topography, does not go with the infrastructure group. Irrigated agriculture has not been influential in promoting urbanisation. Overall, one finds disjointedness in urbanisation and development; this is due to ad hocism. The efforts have fructified in the development of quality infrastructure in less developed areas, which has facilitated rural-urban interaction and thereby changed rural–urban dynamics. Mukherji and Phadke suggest that the authorities should take advantage of this opportunity by developing a proper perspective and rural orientation to make the community there feel that it is an equally respected partner in the development process.

The first paper in the fifth group, '**Decolonisation of the Administrative Organisation of Space in Punjab since Independence**' by **Gopal Krishan** and **Surya Kant**, seriously and critically reviews the process and pattern of the phenomenon of administrative re-organisation of space in Punjab during the post-Independence period. There is a persistent demand for new administrative units like states, districts, etc., with the premise that smaller, compact areas lead to healthier and people oriented functioning. This is despite the fact that the benefits of such efforts are often restricted to the local elite and the process of division continues unabated. The authors have brought out the diametrically opposite trends prevailing in developed and developing countries. They have placed their analysis in a historical context to show how the philosophy behind re-organisation has been different during the colonial and postcolonial periods. The developments during the post-colonial period have not always been well intended. In the initial stages, the authors point out, re-organisation was used to bring about administrative reforms, but later, there were excesses; populism has further aggravated the situation. The process becomes complex because there is a multiplicity of divisions due to the requirements of different agencies, whose jurisdictions often do not coincide. Gopal Krishan and Surya Kant also bring out the diverse pressures that are reflected in the varied sizes of units even at the same hierarchical levels. They, therefore, raise three important issues at the end viz. the size and shape of the units (districts); the methodology of their delimitation, and the number of tiers to be employed. The authors are well aware of the fact that these aspects cannot be identical in all areas. But a certain uniformity in basic principles becomes essential, they contend. Moreover, they rightly argue that multiplicity of layers often involves delays and the number of tiers need to be predetermined.

In the next paper, **'Development Problems of Rural Settlements in Difficult Ecological Environments: Cases from Gujarat'**, **Anjana Desai** raises issues related to the sustainability of settlements in three backward regions in the hinterland of Ahmedabad city, Bhal, Nal and Panchal. Desai examines rural sustainability in terms of physical parameters vis-à-vis socioeconomic and demographic pressures before looking at the feasibility of alternative development inputs in the form of activities for the economic stability of the people inhabiting the concerned regions. The paper is a fine example of adopting modern technology for putting into operation traditional concepts such as land capability and suitability. The selection of micro regions has enabled Desai to make empirical observations for the regions from very close quarters and get a feel of the regional problems. The policy prescriptions are, therefore, realistic and workable. In the period when the emphasis in research is on problems related to urban-industrial areas, especially by academicians from metropolitan cities, Desai has chosen to work on rural areas with a degraded agricultural environment. She has effectively focused on issues related to the sustainability of the rural economy and she firmly believes that the problems of urban sustainability cannot be bifurcated from those of rural economic stability. Desai emphasises that self-sustaining villages with basic amenities alone will check rural-urban migration. She is optimistic that the creation of a suitable rural environment would attract city dwellers to the peaceful, rural surroundings.

The third paper in the group, **'Fluvial Environment and Development of the North Bengal Plains with Special Reference to Land Use Problems'** by **S.C. Mukhopadhyay**, utilises the geographer's skills in interpreting the fluvial environment for development with special reference to land use problems in the North Bengal plains. Mukhopadhyay focuses attention on the need to understand the complexities in the evolution of fluvial land forms, brought in by varied structural and climatic conditions, before taking up any development projects. He emphasises that technology cannot overcome all the constraints related to environment and a careful understanding of physiography and drainage characteristics in relation to geology and climate is required before arriving at a decision to classify land into particular categories. Mukhopadhyay demonstrates that evolutionary episodes of the river basins need to be carefully studied in order to understand the diverse resource potentialities and problems of land use in the area. He brings home the fact that a classification of the area into morphogenetically significant terrain units should precede identification of natural development corridors for various activities, artifacts such as communication routes, urban-industrial complexes, cultivation, etc. Mukhopadhyay's suggestions on solution to problems faced in utilising land in the river basins are commendable.

The paper titled '**Two Ancient Ports of South India – Muziris and Colchi: A Comparison in Maritime Trade**', by **S. Ebeneezer**, is a pertinent example of using knowledge of the home environment to provide a geographical interpretation to historical events. The work has originated in the academic concern of the author about the decline of old port cities in India; the cases of two such ports of South India are cited here. Ebeneezer has drawn attention to the sequence of historical events leading to the emergence of these ports around first millennium B.C. and their subsequent rise to the status of maritime trade emporia. Realising the differences in location with respect to site and situation, the author has attempted a comparison in the maritime trade of the Muziris and the Colchi. While exploring the factors for the decline of these ports, which once possessed a commanding location at the mouth of the rivers, the author has also gone into significant causalities, such as the nautical environment in which the ports had flourished, trade composition and links using archaeological and historical evidence. Ebeneezer also raises the issues related to their differential trade orientation and, thereby, the varying degrees of their prosperity. While the Muziris on the West coast showed trade relations with the Roman empire, the trade of the Colchi on the East coast was conducted primarily with the countries of Southeast Asia. There was also a difference in the composition of the goods exported. The former exported a more diversified package, including spices, ivory, beryl stone and fine textiles, while the latter depended on the export of a sole product, viz. pearls. This resulted in greater prosperity for the former. However, both ports suffered a decline in due course because of natural events like silting of the river courses and shifting of channels and other socioeconomic and cultural factors.

The fifth paper of the group, titled '**A Perspective on Solid Waste Management in Bangalore City: Viable**

Technological Options', by **Daksha Barai**, is a micro level study on the problems of waste disposal in the city of Bangalore. The public sector formal system of waste disposal and management in Bangalore city, which includes a solitary plant at Bommanahalli, is essentially supported by community-based formal systems of garbage management, incorporating research institutes, the GKVK University of Agricultural Science and a number of non-governmental organisations. Barai explains that besides the above, informal systems of waste management also exist in Bangalore, incorporating various producers of recyclable wastes, like households and commercial and institutional establishments that arrange to sell the waste through a hierarchical structure of networks. In the entire framework of these formal and informal systems, recycling enterprises play a significant role. In addition to all these, Barai feels, the inclusion of private sector efforts would be further helpful in making the waste management system of Bangalore more efficient and it would also generate employment in the unorganised sector.

The paper, '**Social Infrastructure: A Study of Housing in Ahmedabad City**', by **K.M. Kulkarni**, dwells at length on housing, which is a very important component of urban infrastructure. Houses are not always available in the required quantity and quality in urban areas and urban dwellers have, therefore, to manage with the available housing facilities. The problem becomes more acute as the city grows in size. Kulkarni's discussion is restricted to three crucial aspects: housing stock, old and modern house types and housing deficit. The spatial patterns of these aspects have also been brought out effectively. The author has related each of these aspects to the physical environment and history of evolution of the city. He has highlighted the paradox of better housing stock with higher deficit and poor quality in one part vis-à-vis low stock but marginal deficit and better quality in the other. He has also brought out the difference between the old and newly settled areas. In the same context, Kulkarni raises the issues related to crowding in cities and the vulnerable sections of urban society who suffer from such crowding. At the end, he provides some hints for future planning of housing stock in the city.

The first article of the sixth group, **'Perspectives on the North Konkan Vegetal Cover'** by **Sudha Srivastava,** addresses some of the issues related to depletion of vegetal cover in North Konkan. Srivastava expresses concern over the rampant deforestation in the area initiated by the metropolitan expansion of Mumbai since the colonial times that took advantage of the access provided by the creeks penetrating the land. Based on the analysis of data compiled over a long period, she establishes that the land, which was rich in tropical deciduous or semi-evergreen vegetation once upon a time, has suffered at human hands during the colonial period because of maritime trade. Highlighting the replacement of natural vegetation by commercially valuable species, Srivastava also examines the damage caused to teak trees by the introduction of the railways. The author brings out vividly the crisis situation reached because of the progress of urbanisation around Mumbai, which has forced the administration to make room for built-up space. Srivastava has taken note of the crucial role played by the Forest Department in protecting the forest cover, but points out some limitations in the functioning of the department as it has virtually no control over urban development; she also laments the political and administrative exigencies in regularising encroachments. Srivastava has drawn attention to certain innovative policy measures taken up recently and hopes that they will provide some protection to the forest cover.

The next article, **'Distribution of Agricultural Land among Social Groups in Western Maharashtra'** by **A.S. Jadhav**, tries to focus on a scenario of social disparities in the rural areas of the Western Maharashtra plateau. Jadhav brings out a critical issue related to injustice in the distribution of agricultural land among social groups. Using empirical observations as well as primary data collected from the households, Jadhav emphasises that the disparity is not restricted to the availability of agricultural land, there are also wide differences in the quality and productivity of such land. The issue has been examined in relation to the land/man ratio among the different groups to bring out the fact that some communities possess a historical advantage in holding such land, which is not even cultivated by them in some cases; this is even more painful than non-availability. Jadhav explains how an initial advantage is translated into a dualistic occupational pattern, leading to differential empowerment. It also produces gaps in the affordability to use necessary technology and take advantage of various development schemes introduced for the benefit of the farm community. One thus finds the synergism effect leading to magnification of the initial crisis. Jadhav rightly laments that even after fifty years of independence, socioeconomic disparities in land ownership remain pronounced despite the various policies adopted and the Acts enacted to bring about a change in the pattern of land ownership.

The three articles in the seventh group concentrate on the economic, demographic and development aspects of Mumbai and its metropolitan region. In the first article, **'Functional Change and Population Displacement within the Core Areas of Greater Bombay'**, **Heinz Nissel** historically traces the evolution of the two distinctive core areas in the southern part of the city of Mumbai, namely the Fort Area, and the Bazaar Zone. An outcome of the interaction of various social and political groups, including the colonial rulers and the indigenous communities, these two zones lost population in later years as a part of the city expansion process. In minutest detail, Nissel examines the ward-wise population loss in the above core areas during 1961-91 and its relation to the pattern of population density. Apart from the core areas, Nissel argues, loss of population has also occurred in the Central Cotton Textile Mill areas with a simultaneous increase in the North-central parts in Sion and Mahim, located at the extreme north of the old city. Nissel's primary explanation for the above process centres around the competition for urban land between residential and commercial activities in which he finds the dominance of commerce and business over residence in areas like Nariman Point or Cuffe Parade. While identifying the existence of community neighbourhoods according to religion and language in areas surrounding the Bazaar Zone, Nissel talks about the decline of the erstwhile interactive space and the subsequent growth of segregated space in the residential areas of these parts of the city. He also finds in these 'leftover' residential quarters of the Bazaar Zone, a sinking urban space, containing a mixture of residence and commerce, which is poorly maintained. Exposition of the above space links Nissel's paper with various contemporary analyses on the newly constructed marginalised residential space of the poor in the global cities in different parts of the world.

In the next article, **'Manufacturing Workers in the Informal Sector: Typology and Organisation of *Jari* Workers in Mumbai', Smita Gandhi and Swapna Banerjee-Guha** examine the life and work of a marginalised community of Mumbai, the *jari* workers, in the wake of the liberalisation era in the 1980s. 'Jardozi', as the embroidery work is popularly known, is an old traditional craft in Mumbai. Although it suffered a distinct setback during Mumbai's industrial expansion phase, it never became extinct from the city's economic base. The reason for it, as argued by Gandhi and Banerjee-Guha, is its tremendous adaptability to the changing market and lifestyle of the consumers, the latter undergoing considerable change in location and diversification. All these together, the authors suggest, created a reconstructed economic space in the city in specific locations with the mushrooming of small *jari* workshops in newly developed slums and low class, Muslim dominated residential areas. In other words, quality of life and community identity, intertwined with this activity, were quite focused. The detailed analysis of the social and economic life of Mumbai's *jari* workers finally focuses on the plight of traditional craft workers in Indian metropolises in general. Unlike many other craft based workers with less adaptive capacities, such as potters or broom makers, although the *jari* workers have tried to diversify and opt out for mass production, in Mumbai, they face a perpetual lowering of wage and increasing fragmentation of work, thanks to the perennial flow of rural Muslim inmigrants from the depressed areas of Uttar Pradesh and Bihar into Mumbai. Their status, thus, is largely that of a socially and spatially marginalised group.

The third article in the group, '**Urbanisation and Related Issues in the Mumbai Metropolitan Region, 1961-2001**', authored by **V.S. Phadke**, in comparison, deals with the wider issues of urbanisation and development in the Mumbai Metropolitan Region. His basic concern is the changing relationship between city and its region, in the present case between Mumbai and its metropolitan region, from 1961 to the end of the 20th century, including the initial years of globalisation. While examining the process of growth, Phadke tracks down the trajectory of urbanisation of the metropolitan region of countries like India, where the initial boost came from trade, commerce and manufacturing – the basic economic sectors – and, later, from the initiation of globalisation through the communication, construction and service sectors. The crucial developmental problems of the metropolitan regions in the contemporary era, their gradual de-linking from a symbiotic relationship with the metropolis and the resultant economic decline are reflected in this analysis. In the Mumbai Metropolitan Region, employment in trade and commerce has gone down, while manufacturing is on the decline. A large part of the urban growth in the region in the post-1980s period has occurred due to the sprawl outside Mumbai into the hitherto green rural areas, which has led to a disruption of both the environment and the society. Phadke critically views the city centred development plans of the recent years at the cost of marginalisation of the surrounding rural areas. The study thus gets linked up with a contemporary debate of development studies, the urban bias in regional development.

We want to place on record the immense cooperation and help received during the course of the preparation of this volume. In the first place, we would like to acknowledge with thanks the contributions made by individual authors to support the venture, as without them, it would have been impossible to bring out a book of this type. We would also like to place on record our appreciation of the help rendered by G.P. Mahulikar, head, Department of Sanskrit, University of Mumbai, for helping us in inserting diacritic marks during the transcription of Sanskrit verses. Our thanks are due to Prasad Gogate, Jean Edge, Ravindra Gawde and Sonali Redkar, who helped us prepare a draft of papers from the original manuscripts. V.S. Rao, Dipti Mukherji, Prasad Gogate and Ravindra Gawde have also rendered help in preparation of the final maps and we would like to acknowledge their help too. Not the least important is the assistance provided by members of the faculty and non-teaching staff. We appreciate their indulgence. Last but not least, significant are the efforts put in by M/s Rawat Publications in bringing out this volume in a professional manner in record time. We appreciate their dedication in making the volume a reality.

V.S. Phadke
S. Banerjee-Guha

1

Environment, Sustainable Development and Tradition

S.P. Das Gupta

Introduction

It goes without saying that no economic developmental effort can be sustained long enough under the current system of massive destruction and wanton exploitation of scarce natural resources, including both renewable and non-renewable ones. It is virtually impossible to achieve sustainable development of the economy without paying adequate attention to the need to maintain the vital ecological balance between all elements of an environmental set-up. The process of development can be fruitful only if all the available natural and human resources are wisely and thriftily used, rigorously conserved, rapidly regenerated and thoroughly renovated, wherever possible, without upsetting the currently prevailing ecological balance within a geographical region.

According to the traditional system of economic development, sustainable growth entails continuous production of wealth along with pari passu conservation of natural and human resources as per the ancient Indian doctrine of *yoga-ksema*. While modern science and more efficient technology can only lead to higher production levels, so also sustainable growth and development of an economic process can only be achieved by adopting the traditional view of production-cum-conservation of the available natural and cultural endowments and equitable distribution of the same for the welfare of all.

According to the deterministic school of thought in human geography, a community or a social group, traditionally occupying a certain well-defined territorial space, usually depends upon the environmental resources available from the surroundings and upon the constraints prevailing within its habitats for its sustenance. The material cultures, comprising agricultural practices, indigenous industries, handicraft products, art expressions, technological attainments, etc., of social groups are intimately related to the set of environmental parameters within which a society is placed as a functional entity. It is said that as human groups continuously proceed on adapting to the prevailing environmental set-up, their material culture, which consists of articulating crafts, economic products of different kinds and social institutions, goes on developing systematically. In any case, in a given spatial organisational layout and social fabric, the symbiosis between human species and the other members of the natural environment is generally achieved after long-continued interaction between man and environment, leading to the evolution of a set of well-established traditions and ethos along with the development of technological acumen, art forms, ethnic crafts and agro-industrial products.

Habitat, economy and society are intimately related to each other in an extremely complex format peculiar to all higher order organic systems. Our habitat, with all its environmental nuances, coupled with ecological checks and balances, supports and nourishes the economy and maintains the health of our society. Similarly, at the same time, the extant society, too, through its tentacles of diverse economic and social actions, must necessarily have to protect and nurture the entire environment prevailing with its habitat; otherwise, the society cannot sustain itself long enough, nor can the habitat remain habitable for a great length of time. A successful society is prone to follow a natural growth path opened up by a set of technological attainments, cultural traditions and ethos, which have developed within the limiting conditions prevailing in an environment. Therefore, to go in for adoption of any new developmental alternative in any area, one has to look deep into the already established traditions first, with a view to striking at the most suitable or the most appropriate technological option for bringing about a planned growth of the culture and economy of a social group.

It goes without saying that a society cannot remain static for any length of time. According to the fundamental law of nature, everything in this phenomenal world of ours must undergo change along with the passage of time and with the shifts occurring within the spatial organisations in different parts of the globe. But, for a really sustainable growth of culture and economy, the society has to carefully regulate the change to ensure that the desired level of social and economic development takes place along the right direction as generally set forth through local logical ingenuity and long-standing traditions. The keyword for economic and cultural growth of a society lies in deliberately building up a truly benign habitat with all its diverse member species properly fitted into the entire milieu in a sort of organic holistic system. The main point here is that while the environment undoubtedly nourishes our societies within the given habitats, the human social organisations also have the bounden duty, not only to conserve and protect the available environmental resources, but also to enhance the same so as to make the habitat an increasingly better place to inhabit for the human as well as all other member species through the years to come.

Traditional Backdrop

But what is there in the so-called tradition which we are supposed to value and preserve for attaining a substantial growth of the material culture and social economy? In any case, how does a human tradition really operate within a society in fashioning the future in a sustainable manner? The value of well-tested and long-standing traditions is immense, especially in an ancient cultural domain like India. It is well known that tradition dies hard, and it generally seeps down into the innermost depths of one's very existence, as it were. However ultra modern a person may be in terms of economic, scientific and technological advancements, he cannot easily escape from his own cultural and emotional backdrop, which generally goes, as though, into the very gene structure of a person or a community. Thus, knowingly or unwittingly, the growth of economy and material culture in a society, along with its art forms and technical production systems, has to take traditional values into active consideration, and a society has to tackle the economic issues and environmental problems of today accordingly.

Take, for example, the serious question of global warming, which is often hitting the newspaper headlines nowadays. Or, let us look at the widespread problem of the severe resource crunch, which is looming large over the entire Third World today. It is now well known that the environmental crisis of global warming has been on a continual rise chiefly due to the fossil fuel based industrial upsurge in the post-industrial Western world. Similarly, notwithstanding the availability of enough natural and human endowments, the greater part of the world has been suffering from stiffening inequity of resource distribution. Overgrowth of the economy, scant regard for human rights and over-consumption of resources accompanied by over-wastage of the world's natural endowments in certain parts of the world, ignoring all traditional wisdom and value systems, are the real causes of the environmental problems of our unequal worlds.

Traditional wisdom never supports environmentally unsound and socially unequal development options. Let us take some insights, for instance, from the traditional ancient Indian approach to sustainable growth of the economy and material culture.

Basic Precept

We must accept, ab initio, the fundamental premise that the common natural resources of the world cannot belong exclusively to any favoured nation or group of people of a particular period of human history. According to the traditional Indian pattern of thinking, one has to exercise extreme self-restraint in enjoying the earth's common endowments, as has been clearly enunciated in our *Vedas* and *Upanisads*. One of the most oft-quoted verses of the *Upanisads* says:

Is a vasyam idam sarvam yat kim ca jad'atyam jagat,
Tena tyaktena bhunjitha ma grdhah kasya svid dhanam.

(Is opanisad, Verse 1)

This means: "Whatever is there in this ever-changing world of ours, which is always in motion, all that has to be considered as being pervaded by a single controlling factor. Therefore, you should nurture yourself with only that portion of the resources which has been showered upon you. You must not covet anything more than that for, after all, to whom does all this natural endowment really belong?" Natural wealth, being a common property, obviously belongs to none else than the nature itself. In his celebrated commentary on the above *Upanisadic* verse, Sankaracarya aptly concludes by urging, "Cherish not the slightest desire for wealth. Do not harbour lust for any resource whatsoever, whether the same belongs to someone else or even if it belongs to your own self. In other words, do not covet at all any wealth because in reality nobody can own a common natural endowment, which one may covet." And then, "With such a spirit of self-restraint and sacrifice" – the *Is opanisad* (Verse 2) hastens to add – "one must go on working hard enough to sustain a full span of the life cycle *(kurvann-eveha karmani jijiviset saten samah).*" The *Upanisad*, therefore, squarely emphasises that no person or group of persons can exclusively and wholly own any common natural resource for their own exclusive use. Truly speaking, natural wealth belongs commonly to all, spread over all times – past, present and future. And people must, therefore, strive hard to develop and conserve all available natural resources for the very survival of the entire society. If this wise principle had been followed, there would not have been any serious disparity, inequality and differential development of the economy and cultural traits as we find today.

But the hard fact is that all our common natural wealth – whether at the local, regional or global levels – has literally been usurped and over-exploited by unholy means over many centuries now. As a result of this unhindered exploitation of common resources, the earth is fast being denuded of its natural endowments, especially those that are non-renewable; in addition, the environment is getting defiled more and more by hazardous chemicals and the waste discharges of modern times. If we cannot cry a halt to this trend right now and find a way out based on abiding wisdom, the world is going to be a totally unlivable planet in the not so distant future.

Resource Generation

But does this mean that in trying to turn the world away from impending environmental disaster and threatening resource crunch, we must cut short all our modern economic activities and revert back to the primitive mode of production and the original way of living? Not at all, for many of the old and outmoded means of production were rather wasteful and uneconomic. When technologies, whether traditional or recent, are driven by the unholy forces of merely market-oriented economy, just to accumulate money and nothing else, the consequential social and environmental effects on the biodiversity are usually disastrous. That is precisely why, in the ultimate analysis, things have been going wrong, as is happening today. In any case, it goes without saying that to sustain and nurture human civilisation, the society must have to produce sufficient amount of material resources, especially those that are renewable, so as to meet the usual social requirements and the minimum needs of everybody.

Unlike the contrary notions harboured by many, the Indian tradition actually favours ample production of wealth, which should, however, be lawfully generated and multiplied through ecologically balanced use of common natural endowments, like land, air, water, energy or power and space. While developing the available natural resources, it has to be ensured that the means of production adopted for the purpose do not inflict any serious or irreparable damage to the earth and to its component members. An *Upanisadic* injunction in this regard clearly enjoins:

Annam na nindyat. Tad vratam.
Annam na paricaksita. Tad vratam.
Annam bahu kurvita. Tad vratam.
Prthivi va annam.

(*Taittiriya Upanisad 3/7-9*)

The purport of the above *Upanisadic* statement is, "Wealth must not be despised at all – that should be the vow. Wealth must not be wasted – that should be the duty. Wealth ought to be developed manifold – that should be the determination. The earth should be treated as wealth by itself."

Thus, it has to be noted that the well-known dictum – 'produce or perish' – held equally good in the traditional pattern of thinking in India as it does in the modern economy. But this wealth has to be produced without inflicting any injury to Mother Earth, unlike what is going on now. After all, the earth is our only available and ultimate resort. Obviously, in the present circumstances, modern science and scientific innovations of the appropriate category can only offer the best technological options that must be adopted as the most viable means of production of resources meant for common consumption. But the technology so used must have to be environmentally sound and eco-friendly, yet humane enough to be free from personal greed of any kind as per the tenets of traditional wisdom enshrined in the *Vedas*, *Upanisads* and in the ancient Indian *Dharma Sastras.*

Sustainable Development

The basic principle of sustainable development – that is, generation of all useful material resources, parallel with conservation and enhancement of the natural habitat – has been nicely put in the *Mahabharata* and in the traditional *Smrti* texts or the law books of India. One such well-known eco-friendly policy statement is:

> *Puspam puspam vicinvita mulacchedam nakarayet,*
> *Malakara ivarame na yathan gara-karakah.*
>
> *(Mahabharata 5/34/18)*

The verse says, "Resources should not be generated by extracting produce with great discretion, as is usually done by a florist picking and choosing from blossom to blossom with the utmost care in a garden, without uprooting the plants, unlike the charcoal maker", who mindlessly fells and burns down young trees for the production of wood charcoal. A parallel statement also occurs in the *Parasara-Smrti* (1/62). Thus, Indian tradition insists that resources should be generated with tender care without causing any imbalance in the ecology. This principle has been especially emphasised in the Buddhistic teachings as well. For instance, it has been said:

> *Mrn-mardana-trna-ccheda-rekh adyaphalam-agatam,*
> *Smrtva tathagatim siksam bhitastat-ksanam utsrjet.*
>
> *(Bodhi-caryavatara 5/39)*

The purport of the above verse is, "Having due regard to the teaching (of compassion) advocated by the Buddha, one must immediately stop, out of awe-inspiring reverence, such injurious acts as of forcibly upturning the soil, removing the (green) cover provided by grasses, etc., or for that matter, cutting deep down into the earth." Thus, in the traditional production system, concern for maintaining the ecological balance was paramount, as it should be for human welfare.

Energy Input

The wise people of the Orient also realised that energy being the prime mover of all fruitful actions, provided the real means of efficient production and resource generation. The method of getting access to the sources of energy has been aptly enunciated in a Vedic mantra:

Divas prthivyah paryoja udbhrtam
vansapatibhah paryabhrtam saha,
Apam ojmanam pari gobhir avrtam
Indrasya vajram havisa ratham yaja.

(Rg Veda 6/47/27, Yajur Veda 29/53)

The above mantra enjoins, "May ye, o wise ones, harness all around the energies held fast by the stars and the earth. Collect, all over, the available nutrition inherent in the trees, and make use of all the energy as available in the running waters. Hold on and utilise the power of the thunderbolts of the sky and that of the crafts flooded with rays." Thus, in the *Vedas*, greater emphasis was laid on the use of renewable and non-polluting energy resources like hydroelectric power or solar energy, as far as possible for economic development. All this simply indicates that the traditional attitude towards resource development was a positive and not a retrograde one.

In generating adequate wealth, our dependence on the available natural renewable resources of the inviolable earth and on the clean and pollution free solar power has been duly emphasised in many a *Vedic* mantra. One such mantra is:

Anagaso aditaye devasya savituh save,
Visva vamani dhimahi.

(Rg Veda 5/82/6)

The mantra means, "May we, with the help of the bright solar energy, be free from all wrongs committed against the inviolable earth, so that wealth can be produced." A similar attitude has been expressed in many other Vedic verses as well, such as:

Vajasya nu prasave mataram mahim
aditim nama vacasa kar amahe,
Yasyam idam visvam bhuvanam avivesa
tasyam no devah savita dharma savisat.

(Yajur Veda 18/30)

Here the prayer indicates: "To gain wealth, we celebrate with praises the earth, the mighty mother and inviolable one, on whom this entire living world has settled. May the solar energy promote our habitations here."

Doctrine of *Yoga-kshema*

According to the traditional system, due stress has been laid not only on raising the production level as much as is required, but also on the need to conserve resources and distribute the proceeds thereof in a proper fashion. The production-cum-conservation and distribution nexus has been wonderfully brought to focus in a charming Vedic prayer:

Nikame-nikame nah parjanyo varsatu phalavatyo
na osadhayah pacyantam yoga-kshemo nah kalpat am.

(Yajur Veda 22/22)

This mantra urges, "Let the clouds shower upon us rain waters in plenty and at the right times; may our plants grow in great abundance, laden with ripening fruits; may we win enough treasures not acquired so far, and conserve the resources properly for the well-being of all of us."

The abovementioned Vedic spirit has also been appropriately incorporated in the codes of traditional Indian law books; one such code is:

Alabdham caiva lipseta labdham rakset prayatnatah,
Raksitam vardhayeccaiva vrddham patres niksipet.?

(Manu Smrti 7/99)

The purport of the above instruction is, "One must strive hard to generate sufficient wealth not acquired so far and also to conserve the available resources with utmost care. Whatever resource is thus conserved has to be developed further and whatever wealth is produced therefrom should then be properly distributed to deserving persons and to the rightful places." Precisely, this is what is known in Indian ethics as the doctrine of *yoga-kshema*, that is, the principle of 'development-cum-conservation of resources'. In the same vein comes the oft-quoted phrase:

Yoga-kshemam vahamyaham.

(Bhagavad Gita 9/22)

The main purport of this declaration is, "Resource development-cum-conservation comes to those who follow the path of righteousness."

It must be noted at the same time that along with production of consumer goods in sufficient quantities, the need for equitable and rightful distribution of the produce got equal importance in our traditional social ethics. That is why it has been ordained categorically:

Annadyadeh samvibhago bhutebhyasca yatharthatah,
Tesvatma-devata-buddhih sutaram nrusu pandava.

(Srimad Bhagavata 7/11/10)

The *Bhagavata* says here, "Wealth should have to be equitably distributed among all people. And for that matter, one must treat all persons as being the same as one's own self and as equivalent to the most venerable ones." It goes without saying that the world climate would certainly have been much better if we had followed the above dictum of equitable distribution of resources based on the sound principle of *yoga-kshema*, that is, development-cum-conservation of resources as has been enunciated in the traditional thoughts of India.

The same doctrine is also applicable even in framing the fiscal policy of a state, for the *Mahabharata* advises:

> *Yatha madhu samadhatte raksan puspani satpadah,*
> *Tadvad arthan manusyebhya adadyad avihimsaya.*
>
> (*Mahabharata* 5/34/17)

The verse states, "Just as the bees procure honey (in bits) conserving the flowers, similarly, revenues should be collected (in small amounts) without causing any injury to the people."

Indigenous Technologies

What should be the role of the available means of production in different sectors of the economy? In principle, production of goods and services has to be optimised in diverse fields of economic activity, be it agriculture, animal husbandry, fishing, mining, trade, commerce, transport or industry. Productivity has to be raised by applying the best available organisations and methods or by using the most appropriate and eco-friendly technology so as not to cause any injury to the members of the ecosystem concerned.

Within the framework of the traditional Indian value systems outlined above, what could be the possible role of the indigenous technologies available within the country? As a production technology, the Indian cottage or village scale industry generally employs rather slow processes and is thus an apparently inefficient or uneconomic activity. But, on the other hand, indigenous industries per se are generally more eco-friendly. Yet, what is most important, being a widespread productive process, going deep into every nook and corner of this vast and varied country of ours, small scale and household industries play an extremely vital role in silently bringing about social equity and justice in a great measure, without adversely affecting the fragile ecosystem. Unlike modern manufacturing industry, the Indian handicraft and other indigenous industries do not usually destroy the environment and make the habitat unclean or unfriendly. Incorporation of all aspects into the value systems of the country would help conserve the scarce natural resources, maintain the environmental quality of the habitats and promote social equilibrium at the very grass-root level by gainfully employing the vast reservoir of otherwise idle manpower lying in the rural backyard of India.

If the industrial West has been apparently grabbing and consuming the natural resources of the world at a fast rate and polluting our land, air, water and space resources with undesirable waste discharges, the unfortunate Third World countries, including India, have been dissipating much of their valuable workforce by not employing the available manpower resources fully to produce adequate material wealth. If the West has to be accused of unethical consumerism leading to an upcoming global environmental disaster at no distant a future, the Third World can be held squarely responsible for not making full use of its great workforce reserve and valuable traditional cultural assets for the purpose of human welfare. Fully sustainable and eco-friendly resource development demands that over-consumption of goods and services should be severely curtailed, the craze for more and more material comforts should be contained and, at the same time, it should entail fuller employment of the vast human workforce in productive activity based on the doctrine of *yoga-ksema* as has been enunciated in ancient India.

The need of the hour is fuller development of traditional skills, along with revival of indigenous handicraft and local industries, satisfying the need for both utility products of common daily use and beautiful creations of finer artistic value. India and other Third World countries, in particular, where indigenous industries are still a vibrant social phenomenon, can possibly show the way to achieve really sustainable resource development without incurring the wrath of diminishing returns and without calling for serious environmental risks.

It goes without saying that indigenous industrial skills can very well coexist with modern industrial giants possibly in a sort of symbiotic relationship between them. India's great handloom weaving industry, which is fed by the yarn manufactured in large spinning mills, is perhaps a unique example of the fruitful combination of a traditional skill with modern factory production systems. If we are determined to make much better use of the now idle manpower in rural and semi-urban India, we can even think once again of reviving, e.g. the hand spinning of yarn, especially during one's spare time, and producing textile goods purely for his own consumption at the village level. It is well known that in the great rural backyard of India, a huge workforce lies unutilised as an active resource for a large part of the year. This large idle force should be fruitfully utilised in organised productive programmes, be it turning the spinning jenny or flying the shuttle on the hand loom, or any other useful handicraft making work within its own homes. It is true that the spinning wheel or the hand loom or any other craft technology, for that matter, cannot truly compete with modern industrial giants either in terms of productivity or cost of production in the usual market-economic sense. But if the accrued social benefits and environmental costs are taken into account, the traditional industries would come out in a different perspective altogether. What is needed is a little bit of soul searching and a determined thrust to fruitfully develop indigenous industries – perhaps aided by modern technology of appropriate character – wherever possible.

Conclusion

What is the way out from the mounting environmental crisis of modern times brought about by our intense greed and anti-ecological lifestyles? And then, how to get over the ever-widening chasm between the haves and the have-nots in the world? Possibly, the most viable answer lies in vigorously promoting and improving upon eco-friendly traditional skills in different sectors of the economy for the growth and sustenance of the human civilisation, based squarely on the great message of *yoga-kshema* of the *Upanisads*.

It can be added by way of conclusion that if organised properly and managed with professional skill, the indigenous production technologies, based on the wonderful leitmotifs enshrined in the traditional wisdoms, can indeed assume increasing relevance in bringing about economic and cultural resurgence in the new, brave, healthy world of tomorrow.

2

Managing Environmental Crisis

Parbati K. Sircar

Introduction

Entitled 'Warning to Humanity', a petition sponsored by the Union of Concerned Scientists and signed by 1,500 scientists (including 100 Nobel Laureates), was circulated in November 1992 among heads of government throughout the world. It made a fervent plea for more efficient use of resources, an end to activities that damage the environment, elimination of poverty and acquiring control by women over their own reproductive decisions.

There were ample reasons behind this appeal, for *The Encyclopaedia Britannica 1994 Book of the Year* enumerated several environmental issues of great urgency: climate change, ozone depletion, acid rain, toxic wastes, marine pollution, etc. It is seen that environmental problems range from global to national and, therefore, managing them will require different kinds of strategies.

This article initially focuses on some of the more critical global environmental issues and the efforts to mitigate or eliminate them. Secondly, it discusses the environmental crisis that India faces and analyses the causes and consequences. In conclusion, a few general principles and problematics that emerge from the discourse are brought out.

Global Environmental Crises

Several General Circulation Models predict important climate changes due to increased greenhouse gases (Silver and Defries 1991).[1] Of these changes, there is a 90 per cent probability of occurrence of global mean surface warming in the range of 1.5° to 4.5°C for a doubling of atmospheric CO_2 (or its radiative equivalent from all of the greenhouse gases), global mean precipitation increase, Northern Polar surface warming and reduction of sea ice. There is a 67 per cent chance of northern high latitude precipitation increase, summer continental dryness/warming and rise in the global mean sea level.

Global Warming

Next to water vapour, CO_2 is the most plentiful and effective greenhouse gas. Concentration of CO_2 in the atmosphere has increased by about 25 per cent since the Industrial Revolution and is currently (c. 1990) increasing by about 4 per cent each year. Trace greenhouse gases like methane (CH_4), chlorofluorocarbons (CFCs), nitrous oxide (N_2O), and tropospheric ozone (O_3) are also increasing in the atmosphere. Molecule for molecule, they absorb infrared radiation much more effectively than CO_2 does. Their combined effect is likely to be equal to or greater than that of CO_2 alone. Some of these have a long residence time in the atmosphere.

Ozone Depletion

In 1985, Joseph Farman of the British Meteorological Survey and his colleagues reported in *Nature* that the concentration of stratospheric ozone above Antarctica had plunged more than 40 per cent from the 1960s baseline levels during October. By 1987, it was learnt that the ozone layer over Antarctica had been reduced by more than 50 per cent of the values recorded in 1979, and that at altitudes of between 15 and 20 km, depletion was as great as 95 per cent.

In 1974, it was hypothesised that increasing concentration of chlorofluorocarbons was responsible for the formation of the 'ozone hole'. The depletion of ozone is very effective when high thin clouds, the polar stratospheric clouds (PSCs), are formed in the southern spring. As this will cause more ultra-violet radiation to reach the lower layers of the troposphere, it will affect the human immune system, causing cataracts and skin cancer. Many plants like soyabean, pea and melon are also vulnerable to ozone depletion.

Acid Rain

Nearly all of the acidification resulting environmental damage comes from sulphur dioxide (SO_2) and nitrogen oxides (NO_x) released as gases when fossil fuels, especially coal, are burnt. In the atmosphere, these oxides are converted into sulphuric acid (H_2SO_4) and nitric acid (HNO_3) and then into nitrates and sulphates. The acid droplets form a part of precipitation and come down as acid rain. The pollutants can remain in the atmosphere from a few hours to several weeks and can be driven by winds to distant parts of the atmosphere. Hence, acid rain is liable to cause international friction. When acid rain percolates through the soil, excess hydrogen replaces nutrients such as potassium (K), magnesium (Mg) and calcium (Ca) and can affect plant growth. When the hydrogen concentration increases, aluminium (Al) may be released, which is toxic to plants.

Acid rain is destructive to lake ecosystems as well as forests resulting in forest dieback or *waldersterben* (forest death).

Global Environmental Action

From all accounts, human occupancy of the earth over the last several millennia has resulted in an environmental crisis of a grave dimension. However, over the last two and a half decades, there has been an increasing awareness of global environmental problems and the realisation that there is an urgent need to protect, conserve and manage the environment for sustainable development. Two important events that changed the general outlook were the United Nations Conference on the Human Environment at Stockholm in 1972 and the preparatory meeting at Founex (United Nations, 1971).

United Nations Declaration on the Human Environment, June 1972

The declaration affirmed that in the developing countries, most environmental problems were caused by underdevelopment, whereas in the industrialised countries, they were generally related to industrialisation and technological development. The conference also declared that the natural growth of population continuously presented problems for the preservation of the environment and required the adoption of appropriate policies and measures (Massey, 1993).

Among the 26 principles laid down by the conference, four can be selected to indicate the tone of the document. They are as follows:

No. 9: Environmental deficiencies because of underdevelopment could best be remedied by accelerated development through financial and technological transfer.

No. 12: International technical and financial assistance should be made available to developing nations to include environmental safeguards in their development planning.

No. 16: Demographic policies are required where population growth is likely to place the adverse effects of development on the environment.

No. 20: Free flow of up-to-date scientific information, expertise and environmental technologies to developing countries is recommended.

Convention for the Protection of Ozone Layer, Vienna, 1985, and Montreal Protocol, 1987

Under the auspices of the United Nations Environment Programme, four international scientific conferences were held on the 'ozone hole'. Developing countries were anxious that the agreement should not impede their development. As a whole, they utilise 16 per cent of the world's CFCs, while India and China, with one-third of the world population, use only 2 per cent between them. The protocol was modified to accommodate their concern. The September 1987 meeting in Montreal saw an agreement on the Protocol, signed by 21 nations and countries of the European Economic Community. The Protocol provided an incentive for further cutbacks by industries and for the development of substitutes.

World Commission on Environment and Development

Gro Harlem Brundtland chaired an independent commission pursuant to a resolution approved in 1983 by the UN General Assembly. The Commission was to formulate a global agenda for change. Specifically, the Commission was to propose long-term environmental strategies for sustainable development by 2000 and beyond and to consider ways and means by which the international community could deal more effectively with environmental concerns. The Report, *Our Common Future*, was presented to the General Assembly in 1987.

In the overview, 'From One Earth to One World', the Commission redefined its mandate to re-examine critical environment and development issues and to formulate realistic proposals for dealing with them; to propose new forms of international cooperation on these issues that would influence policies and events in the direction of needed change, and to raise the levels of understanding and commitment to the action of individuals, voluntary agencies, businesses, institutes and governments. The Commission focused its attention on areas of population, food security, loss of species and genetic resources, energy, industry and human settlements, – realising that all of these are connected and cannot be treated in isolation. The Commission's Tokyo Declaration on 27 February 1987 enjoined that all the nations of the world, both jointly and individually, should integrate sustainable development into their goals and adopt the following principles to guide their policy actions: revive growth; change the quality of growth; conserve and enhance the resource base; ensure a sustainable level of population; reorient technology and manage risks; integrate environment and economics in decision making and strengthen international cooperation.

UN Conference on Environment and Development, 1992, in Rio de Janeiro

Also called The Earth Summit, this conference led to several documents: the Rio Declaration, Agenda 21, Convention on Climate Change and, Convention on Biodiversity. The Rio Declaration reaffirmed the Declaration of the 1972 Stockholm Conference and set the goal of establishing a new global partnership through the creation of new levels of cooperation among states, key sections of societies and people, by working towards international agreements which respect the interests of all and protect the integrity of the global environment and developmental system. Towards this end, it propounded 27 principles. Of these, principles 14 and 15 are of great import for developing nations. Principle 14 states that countries should effectively cooperate to discourage or prevent the relocation and transfer to other countries of any activities or substances that cause severe environmental degradation or are found to be harmful to human health. Principle 15 categorically states that where there are threats of serious or irreversible damage, lack of full scientific certainty shall not be used as a reason for postponing cost-effective measures to prevent environmental degradation.

Agenda 21 was devoted to spelling out the problems of and policies for international cooperation to accelerate sustainable development in developing countries. It underlined the handicaps from which developing countries suffer in their trade and financial transactions with developed countries. Developing countries are weighed down with external indebtedness, inadequate development finance, barriers to Northern markets and falling prices for the commodities which dominate many economies in terms of production, employment and export earnings. The Agenda recommends that the Integrated Environmental and Economic Accounting (IEEA) system be established in all countries. Better ways should be developed for measuring the value of natural resources and other services provided by the environment.

UN Convention on Climate Change, New York, 1992

Various resolutions were approved by the UN General Assembly in 1988, 1989, 1990 and 1991 on the protection of global climate for the present and future generations of humanity. The New York Convention had the ultimate objective of achieving stabilisation of greenhouse gas concentrations in the atmosphere at a level that would prevent dangerous anthropogenic interference with the climate system. Such a level should be achieved within a time-frame sufficient to allow ecosystems to adapt naturally to climate change, to ensure that food production is not threatened and to enable economic development to proceed in a sustainable manner.

UN Convention on Biodiversity, 1992

The objectives of this Convention are the conservation of biological diversity, sustainable use of its components and fair and equitable sharing of the benefits arising out of the utilisation of genetic resources, including appropriate access to genetic resources and transfer of relevant technologies, taking into account all rights over these resources and by appropriate funding. Article 15 of the Convention, Access to Genetic

Resources, suggested that each contracting party should endeavour to create conditions to facilitate access to genetic resources for environmentally sound use by other contracting parties and not to impose restrictions that run counter to the objectives of this Convention. There was also a non-legally binding authoritative statement of principles for a global consensus on the management, conservation and sustainable development of all types of forests.

A Critique of Global Action

The foregoing bears ample testimony to the manifold ways in which the United Nations and its agencies as well as an independent commission have attempted to control global environmental problems. Conferences, conventions, protocols and resolutions have sought to contain the damage to the biosphere resulting from human activities. Some of them have borne positive results, e.g. the reduction of ozone-depleting substances. On the other hand, there remain quite a few unresolved issues. Although the United Nations is a world forum for international cooperation, member states have a tendency to look at the problems from a national point of view. Domestic compulsions make it difficult to make objective judgements on global environmental issues. Big power rivalry, the perennial North-South conflict, especially before the dismemberment of the U.S.S.R., the tilt of world financial institutions towards the U.S.A., reneging on commitments made at the global forum, especially with regard to the flow of finance and transfer of technology to developing nations, the adversarial stance of transnationals toward economically and politically weak nations, setting up of polluting industries in many countries in Africa, Asia and Latin America, and dumping of toxic wastes in those countries and their territorial waters – all these have thwarted the goals set up by the United Nations.

Set against this, the independent Brundtland Commission has pleaded the case of developing nations in the world forum and has shown the interrelatedness of factors like poverty, resources and the environment. The Tokyo Declaration of the Commission's guidelines for sustainable development should be seriously considered by all nations to bring about an equitable and harmonious world order.

The Indian Scene

Environmental problems in India may be classified under three categories:[3]

1. A number of problems are related to the poverty of the masses and their inability to have access to appropriate kinds of resources in adequate quantities. As there is the effluence of affluence, so also is there the pollution of poverty. One results from an over-consumption of resources, the other is the outcome of having to do with inferior types of resources in insufficient quantities.

   There is thus a vicious circle of economic impoverishment and ecological degradation. It should, however, be mentioned that poverty exists, not because of an overall scarcity of resources, but rather from the maldistribution of and uneven control over resources.
2. A second group of environmental problems arises out of the pressure of an increasing population on the limited available resources, forcing people in many cases to use marginal resources, thereby causing damage to the environment. These problems are intensified by the poor management of resources.
3. The third category includes all those environmental problems that are consequent upon rapid industrialisation and urbanisation without adequate safeguards and regulatory control. Until recently, there was little or no restriction placed upon the discharge of pollutant particulate matter, hazardous effluents and toxic chemicals in the air, water and soil. The inevitable result has been an alarming extent of environmental degradation.

If the results of several decades of development planning have been unenviable, its environmental consequences have been disastrous. The evidence is all around to see and experience – deforestation, soil erosion, desertification, waterlogging, salinisation, alkalisation, air and water pollution, thermal and radioactive pollution, loss of gene pools and so many others.

The reasons behind environmental degradation and pollution are many, complex and interrelated. Our attitude towards nature and the environment and the desire to dominate them, and our ignorance, arrogance and defiance of the logic of the environment are responsible in a large measure for the predicament we are facing. There is a general apathy among the members of the public regarding environmental matters. A laissez faire economy, a pattern of development which is capital intensive, using inordinate amounts of power and high technology, are all apt to take a heavy toll of environmental resources. Add to this the consequence of the increasing pressure of population on limited resources and of the revolution of rising expectations. Absence of a comprehensive environmental policy, buttressed by effective and enforceable legislation and appropriate organisational and institutional support, and lack of coordination among concerned government ministries and departments have compounded India's environmental crisis.

Recent Positive Changes

It is interesting to note that positive environmental action has come from two different quarters – the Supreme Court and the people at large. The apex court's direction to the Calcutta High Court to form a special division bench dealing with petitions concerning pollution is the first of its kind in the country (the 'Green Bench'). There is a vast pool of experience among our rural folk regarding environmental protection. This has served them well in the past, but changed political, economic and social circumstances have all but taken away their traditional rights and power over environmental resources. About 35 per cent of India's population directly depends on forest resources for food, fuel, fodder, shelter and income.

A resurgence has taken place to win back these rights. For example, in Sambalpur district of Orissa, some villagers undertook active measures for protection and conservation of forests way back in the 1940s and 1950s (Pattnaik and Brahmachari, 1996). These efforts became widespread in the late 1970s and 1980s. In Orissa's villages, community level organisations provided a forum to discuss the changes in their life-support bases and to decide on collaboration. Village forest protection committees (VFPCs) were formed in reserved forests under the guidance of the Forest Department. Now these have been extended to non-reserved forests, too. Apex bodies have been formed for better coordination. At present, the villages are protecting 200,000 ha of forests.

An example often cited goes by the name of the Arabari model. Villagers protected 1,272 ha of badly degraded forest within the Arabari Forest Range of Medinipur district, West Bengal. This was actively supported by the forest officials. In return, villagers were given employment in both silvicultural and harvesting operations, 25 per cent of the final harvest, and fuelwood and fodder collection on payment of a nominal fee. By 1983, a previously worthless forest was valued at Rs 125 million. Forest protection committees (FPCs) were started in other areas. As of July 1990, 191,756 ha of forests, especially of degraded *sal* coppice in the districts of Bankura, Medinipur and Purulia, were being protected by 1,611 FPCs. This model, with local variations, has been replicated in 15 other states through Joint Forest Committees (JFCs).

Many more such examples can be cited from different parts of India: the struggle against toxic waste pollution from drug factories in Kolara industrial area in Bidar district, Karnataka; the saving of Gopi Cheruvu (GC), a perennial tank in the Ranga Reddy district, Andhra Pradesh, from the clutches of real estate developers through *shramdan* by villagers under the active leadership of the GC Porata Samiti, and the Aga Khan Rural Support Programme (AKRSP) in tribal dominated Bharuch district, Gujarat, through an integrated development programme with mass participation and especially involving women.

The role of non-government organisations (NGOs), community-based organisations (CBOs), women's groups (*mahila mandals*) and tribal committees (TCs) have also become significant both as pressure groups as well as doers. Gram Vikas Mandals (GVMs) of different types, Pani Panchayats and tree growers' cooperatives are other laudable examples.

Conclusion

George Perkins Marsh, often called the spiritual godfather of the conservation movement, sounded a note of warning towards the end of the nineteenth century: Man has too long forgotten that the earth was given to him for usufruct alone, not for consumption, still less for profligate waste. This warning is equally valid today. As the foregoing has shown, environmental problems, both global and national, have reached a critical stage and their control will require scientific understanding of the crisis along with the political will and moral commitment to contain them.

Some general principles emerge from this discussion. Economics must be integrated with ecology to enable a planning process that can lead us to sustainable development. A bottom-up rather than top-down way of doing things is preferable. It is necessary to involve people in the decision making process and create institutions based on democratic decentralisation. Environment is not a free goods. Use of environmental resources confers benefits, but it involves a cost as well. In a cost-benefit analysis of programmes and projects, this cost should be included. There are now methods available for this exercise. The 'Polluter Pays' principle is now being replaced by making the social cost of pollution so high that polluters find it more economic to control the pollution at source. A problem that requires careful analysis has to do with the allocation of forests to communities as common property resources (CPR). There is a good chance that the forest will be overused and misused and individuals may not have the interest to maintain them properly. This is the tragedy of the commons as pointed out by Hardin (1968). In Indian villages, homogeneity is lacking with regard to caste and class and so equity is bound to take a backseat in reality. It is, therefore, desirable to have joint forest management with community representatives and forest officials working together under well-defined and well-understood rules and regulations.

Going round the streets of Kolkata, I found one day a graffito which sums up the relation between development and the environment:

If we put the environment first
Then development will last.

Notes

1. The data in this section was taken mostly from Silver and Defries.
2. The facts and excerpts are from Massey with additions from other sources.
3. The section leans heavily on my presidential address to the Eastern Geographical Society, Bhubaneswar, 1991.

References

Encyclopaedia Britannica (1994), *Book of the Year*, Chicago.

Hardin, G. (1968): 'The Tragedy of the Commons', *Science* (Washington, D.C.), 162, 1243-1248.

Massey, A.K., (ed.) (1993): *Environment Resource Handbook*, Prakash Book Depot, Bareilly.

Pattnaik, B.K., and Brahmachari, A. (1996): 'Community-based Forest Management Practices: Field Observations from Orissa', *Economic and Political Weekly*, April 13, 968-975.

Silver, C.S., and Defries, R.S. (1991): *One Earth One Future: Our Changing Global Environment*, Affiliated East-West Press Pvt. Ltd., New Delhi.

United Nations (1971): Founex Report, Mouton, Paris.

World Commission on Environment and Development (1987): *Our Common Future* (Brundtland Commission Report), Oxford University Press, Oxford and New York.

3

A Systems Approach to Coastal Zone Management

Satyesh C. Chakraborty

Task of a Manager of Environment

A manager of ecological systems has to honour two basic objectives. In the first instance, he has to approve measures to mitigate the harmful impacts of the designed human actions through any given project upon a specific environment. Secondly, he has to monitor the processes of change obtaining in the given tract as a consequence of the approved project as also independent of it. Though not obligatory, it is desirable that he act for the improvement of the given ecosystem. All these objectives can be confidently attended to if the manager is aware of the systemic attributes of environmental dynamism prevailing in the ecosystem of the given area. The objective of this paper is to introduce the readers to the problematic implied in such a task.

We are concerned in this paper with the management of the environment in coastal zones. Such tracts have complex environment governed by sensitive ecological processes. One may say that these are fragile systems because alterations caused in one part of it leads to a whole lot of changes in the other parts. This is, of course, a general phenomenon. In any other ecological system, the changes are expressed at relatively slower rates. In the coastal zones, the changes come rather quickly and are manifested over wide tracts. The basic reason for this is that the operating force-fields in this type of ecological system are very strong as sustained by wave actions. Gravitation and wind velocity, coupled with the mass of agitated water, impart upon the process a high magnitude of force. The complexities are further enhanced because these force-fields sustain many processes which operate conjointly upon the materials derived from mass wasting of materials as well as the stock of nutrients available at the interface of land and sea, and are poised in a highly interactive state. Our managers have to learn about these interactions to appreciate the nature of dynamism obtaining in the coastal zone.

Structuring the Task

The manager will find many scholars belonging to universities and research institutions competent to describe and define a number of attributes of a given environment. But no single scholar will be competent to define all the attributes because they are the products of training in one or the other subject disciplines, each covering one or the other segments of the total knowledge on environment. Considering that human beings are yet to arrive at any finality in the knowledge of the total environment, the available descriptions and definitions, though appearing partial and segmented, would be the only useable contents of knowledge. This is a constraint that the manager of environment should be prepared to live with at any given point of time.

However, our manager can have a pleasanter state of mind if he tries to mobilise the support of only those scholars whose expertise is palpably related to the manifest features of the environment in his charge. He should avoid calling for assistance from scholars whose subject disciplines bear no relationship to the manifest attributes of the given environment. If this advice is followed, then a meaningful inventory of the relevant attributes of the given environment would be ready for use.

The inventory by itself will be of limited use in discharging all the responsibilities assigned to the manager. Yet it is essential to prepare the inventory. Additionally, he has to integrate these specific attributes into a whole. This work of integration is necessary to define the relationship between the identified attributes of the environment, without which it is impossible to understand the ecological system. Ecology is a subject that intends to define the relationship between the biotic and abiotic elements of the environment. The starting assumption is that all objects are related to all others until proven otherwise.

As with the case of the inventory, the statements on the relationship between the different attributes of the environment, to start with, will remain segmented and incomplete. The manager has to keep himself in readiness to learn more and more about the relationship, but act with the achieved level of understanding at any given moment of decision making. His mistakes may get pointed out by scholars. To admit mistakes arising from transparent disclosures of the scientific basis of his earlier decisions and to rectify those with newer information and arguments is a fundamental prerequisite for a good manager of the environment.

To avoid frequent embarrassment from decisions being challenged and rejected, our manager may try to prepare himself in advance. He has to acquire the ability to assess the relevance of the stated relationships between the identified attributes of the environment. This ability he may acquire by mobilising scholars who are aware of the required methodology. These scholars would generally try to achieve this objective by using the logic of causation as well as the logic of association. Causation can be applied when a fact (i.e., a stated relationship) is tested as valid over time and/or space. Logic of association, on the other hand, only suggests that there might exist an unidentified causal factor, which could account for repeated occurrence of the same set of attributes in the same space over time. Both these methods refer to time and thereby arrive at an understanding of the processes involved in the emergence of a given environment or ecological system, its viability and the conditions under which it will become vulnerable.

No environment or ecosystem is static or unchanging. This is an unavoidable fact of life. Therefore, no environment or ecosystem can remain viable in absolute terms. There are processes inherent in every ecosystem or environment which lead to transformation from one state to the other. For this reason, it is safer not to define any ecosystem as viable and then recommend it to society as a goal to arrive at inevitably. Instead, pragmatic decisions to alter the state of the environment in stages would prove useful. Our manager, wishing to create a healthy environment or ecosystem, can try to understand the inherent processes constituting the given system and not let those get upset by uncalled for actions by human beings.

Appreciation of Environmental Attributes as System Constituents

The observed attributes and their process-bound relationships define systems. A manager of environment should have an enlightened view of the system in his charge. To attain this ability, he has to classify the diverse sets of relationships that the scholars will literally pour onto him. For this purpose he has to have a taxonomic frame to start with. Taxonomy refers to a method for obtaining generic differential as well as hierarchic ordering of observed attributes of a heterogeneous population. Taxonomy is a starting point for constructing paradigms of orderliness from within chaos. We may attempt to construct taxonomy to group the relationship between the components of the environment of a coastal zone.

We assume that our manager already possesses qualitative statements and quantitative measures on the identified attributes of the given environment in a grouped manner. These groups refer to the facts about lithology, hydrology, atmosphere, soil, biota (wild and domesticated varieties), land use, settlements, demography, service infrastructure, economy, developmental aspirations of the people, existing interest group conflicts, existing power groups in the given society, etc. He may also know some of the relationships between pairs or clusters of such attributes. His task now is to unravel the attributes of a dynamic ecological system within the limitations of the data at hand. He may start with the elements that are palpably showing movements. Dynamism of process is nothing but statements on movement and the consequent change in their patterns. Hence, this emphasis.

In the coastal zone, there are two elements of environment which are continually moving in a striking manner. These are the moving air and the moving water. One is an element of the atmosphere and the other, of the barysphere. These two terms, viz. atmosphere and barysphere, are nothing but the headings of groups into which distinct facts have been classified. We need not go into the other facts of these groups at this stage of exploration. It is better to keep the task simple for the purpose of appreciating the basic methodology. What we need to know is that moving air acts upon water to cause the formation of waves. The ecosystem of the coastal zone attains a part of its dynamism because of the waves of sea water propelled by moving wind. These are incessant in occurrence with ever-changing patterns. To map the changing patterns is indeed necessary for an ecologist and, for that matter, for a manager of the environment. But we may, for the present, suspend such an enquiry. Instead, let us see what the waves do to the other elements of the environment.

The impact of the waves is most pervasive upon terra firma or firm land. The waves may crash on land or swish up and down up to a limited distance. The process thus sustained would ultimately produce a gently sloping (exponential in orientation) land on the water front if the relative disposition of land and sea remains unchanged. On the geological time scale, their dispositions have changed either through uplift of the foundation (basement) of the land or through lowering of the sea level as during the glacial period. Any change, slight or otherwise, will restart the process of wave action upon the land to produce another such slope. Therefore, at a given point of observation, one may observe a vertical bluff rising from the landward edge of moving waves to signify an incomplete state of attainment of the process mentioned above.

The waves play another significant role in landscape evolution. Slowly, but persistently, these waves pulverise (through a grinding action) the terrestrial lithology within the operational area. These also produce sorting of particles from coarser to finer grains over space as a function of time. Therefore, at a given point of observation, one may notice shingles constituting the land surface in the wave-affected area, signifying an incomplete state of attainment of the process under consideration.

It has been mentioned that moving air causes the waves to flow. This is how ocean currents form, which do not generally hug the coastline, but are usually a mid-oceanic phenomenon. However, the wind moving consistently parallel or oblique to the coast may produce off-shore currents. In the Bay of Bengal, the monsoon wind system sustains two offshore currents, one flowing in a Northeasterly direction along the Coromandel and the Northern Sircars coast and the other in a Northwesterly direction along the Martban coast. The two currents meet near the estuary of the Matla river and their combined flow moves southward. We have thus a submarine canyon, called the Swath of No Grounds, extending southward from the mouth of the Matla river produced by the combined off-shore currents armed with pulverised materials as traction loads. Bottom materials towards the mid-oceanic area in the Bay of Bengal.

It may so happen that the prevailing direction of wind hits the firm land at an angle. In that case, the waves will also do the same. One end of the wave will hit the land first and the rest will arrive by sliding laterally. Such an orientation of the waves will cause lateral shift of the sorted materials along the direction of the slide. At one end will be left the coarser materials, while the finer ones will drift along the direction of the slide. If there is a natural promontory extending into the sea or a wall or a dock entrance or a pier constructed like that, then shoals will form on one side of it and a steeper slope into the sea will form on the other side. One can guess correctly that the shoal is a result of accumulation of materials drifted by the waves that are oblique to the land. The steepening of the slope of the land on the other side should then be seen as a product of loss of materials through drift. This means that a beach can be maintained only if regular replenishment of drifted materials is assured.

The drifting materials need not always hug the beach. Sometimes, these may extend like a promontory, with a broader end hugging the land, but narrowing along the line of drift oblique to the beach. These are called spits and are the visible expressions of coastal deposition. Sometimes, the seaward end of a spit may turn inward towards the land. Then the entire landform is called a hook. Quite often, a part of the drifting material may get dragged towards the sea and get deposited in the shape of an elongated low wall parallel to the beach. These are called off-shore bars and their locations can be guessed by the waves breaking on them. Where the coast is rising or when the sea level falls, the offshore bars can emerge as large spits, sheltering a lagoon like marine water body on its landward side. In short, the land forms bear testimony to the nature of the operating processes.

It should be mentioned that all wave breakers do not occur only due to the presence of off-shore bars. Breakers are natural expressions of nearly all types of waves and are related to the interaction between wave length and the falling depth of the water while approaching land. Wave length is a function of fetch, which is a product of wind velocity, its consistency over time and the length of water surface traversed. The bigger the fetch, the larger is the wave length. Wave movement is rotatory. The radius of rotation is half the length of the wave. If the approaching wave hits the bottom, where the depth of water is less than the radius of rotation, then the bottom of the wave drags and the top leaps forward to break. The leaping wave is called surf. Along many parts of the Pacific coastline, a lot of surf is seen, primarily because the fetch is generally very large here. Many Japanese painters have been fascinated by the beauty of surf and have immortalised it in their paintings. Apart from these artists, surfing itself is a popular water sport.

Returning to the Bay of Bengal, we may recall that here, the wind system varies between seasons. During summer, wind driven water is pushed against the land. During winter, on the other hand, the wind system tends to push off oceanic water. This can happen only with surface water. But it helps the cooler water from the bottom or deeper parts of the Bay to well up, bringing nutrients on to the surface for the fish. Hence, fishing becomes easier along the Medinipur coast during winter.

Wind is not the only propulsive force that causes movement in sea water. The oceans are sensitive to a much stronger force field, called gravitation, which causes upheavals on oceanic water bodies. This, in common parlance, is referred to as tides and classified as flood tide and ebb tide. During flood (high) tide, the oceanic water rises to invade the land mass. Thus, the land mass at the water front will have a band like zone marked by high water and low water lines. This band of land will be broad if the slope of the land is gentle. Narrower bands are found in tracts with steeper slopes. This band of land is commonly referred to as the beach, and the surface hydrology here will oscillate through transgression of marine and terrestrial domains with well-defined regularity. This tract is, therefore, the most sensitive component of the ecosystem of the coastal zone. In India, such tracts are generally defined as Coastal Regulation Zone I, where no developmental action for economic pursuits is permitted.

In the Coastal Regulation Zone I, the tides should be seen as the primary causal element of the ecosystem. It has already been mentioned how wind-driven waves cause changes in the orientation of the beach. Such elements of the beach landscape should, therefore, be seen as hierarchically subordinate components of the ecosystem produced by tidal action.

The movement of the wind does not stop at the high water line. It operates over a much wider tract. Therefore, the landscape beyond the high water line on the land can get linked up with the ecosystem of the beach through the action of the moving wind. The most pronounced feature of such a linkage is manifested through the formation of coastal dunes. The well grounded and well sorted beach materials, during the period of ebb tide, after drying, get dragged or lifted by the wind moving towards the land. The materials thus dragged or floating get deposited as soon as an obstacle is encountered. The obstacle may be a line of trees acting as wind breakers. Even creepers, shrubs or grasses can halt the movement of beach materials to form dunes. All along the coastal zones of West Bengal we find such dunes, which are decidedly parts of the terrestrial land form, though constructed of beach materials.

In a similar fashion, rivers carry sediments into the sea beyond the beach. These get deposited on the continental slope and can lead to the formation of shoals or islands. In the off-shore zones of West Bengal, we have many such shoals and islands. These are decidedly features of the marine landscape though constructed of terrestrial materials. Thus, the coastal dunes and the off-shore shoals or islands are sub-sets of the coastal zone ecosystem because they are the products of the movements of air and water.

Beyond the Continental Shelf, the true marine ecosystem prevails, though its landward boundary is continually changing its disposition due to the seasonally varying amplitude of the operating forces. For example, during the monsoon floods, the terrestrial contributions extend farther into the sea than during the dry weather period. Similarly, beyond the coastal dunes on the land area, a true terrestrial ecosystem tends to gradually establish itself over the zones of marine influence. It is important to note this conditional statement. This statement means that the terrestrial ecosystem slowly but persistently tries to push out the forces of the marine ecosystem to establish itself.

We may then say that the coastal zone ecosystem is formed of several sub-sets: (a) the Continental Shelf, (b) the beach, (c) coastal dunes, and (d) their tidal zone. All these are interconnected by the forces of moving water and wind. The land forms within each sub-set are visibly distinct.

Building a Comprehensive System

The advantages of landscape analysis should now be obvious. The subject discipline that proves most advantageous to carry out such work is called geomorphology or the science of evolution of land forms. In India, this discipline is commonly taught as a sub-field of geography. In the United States of America, the subject of geomorphology was primarily developed by the geologists. In the erstwhile U.S.S.R., geomorphology was taught by both geologists and geographers. In more recent years, some departments of geology in the universities of India have introduced this subject in their curriculum.

Apart from geographers and geologists, the utility of geomorphology is getting appreciated by many other scientific disciplines. Pedologists or soil scientists are gradually appreciating the utility of this science. The reason is easy to appreciate. Soils can either form in situ or through transportation and deposition of materials. The first form is essentially a product of weathering and the other is a function of erosion and deposition of weathered materials. Both these processes are subjects of study for geomorphologists. Botanists, especially the forestry people, are also using the analytical findings from geomorphology. They have noticed that successive sets of species of trees establish themselves on soils maturing over time. For example, the sandy soils brought in by rivers in spate in the North Bengal Terrai region invite different varieties of grass, to start with. Once these stabilise the newly formed sand beds, then *khair* and *sisso* trees establish themselves. As the *sisso* starts becoming the climax vegetation, the *khair* trees are replaced by *bija* trees. As *bija* becomes the climax vegetation, *sal* trees come in. Geomorphology helps botanists to guess how soon the soil will be stabilised and in what form.

Based on the foundation of geomorphology, pedologists and botanists have developed the science of edaphology. It is expected that all sciences concerning the environment will soon discover the utility of geomorphology because this subject discipline acts as the core to integrate the grouped inventories on the environment prepared by the different subject disciplinarians. It is a system integrator. Some more examples will clarify this observation.

Geomorphologists say that the evolution of a landscape is a function of structure, process and stage. By structure, they refer, as the geologists do, to petrology, mineralogy and their disposition given by tectonics. This is how the inventories on the attributes of environment, prepared by conventional geologists, find a place within geomorphology. By process, they refer to temperature, moisture, precipitation, air and movements. Thus, the inventories on the attributes of environment prepared by climatologists and hydrologists find a place within geomorphology. The interaction between structure and process is conceived of by geomorphologists as a function of time and, other things remaining the same, the temporal changes, they consider, will come to a final state. Hence, the observed landscape is to be seen as a stage in the life-cycle of change. Thus, the epistemology of geomorphology is the same as that of ecology.

Let us elaborate upon the above theme without being exhaustive with reference to some of the attributes of the coastal zone of Medinipur in West Bengal. The petrological and mineralogical attributes are discernible in the land forms produced by moving water and air. For example, lots of base metals, like arsenic, bismuth, mercury, lead, zinc, copper, gold, silver, uranium, etc., are found in the coastal zone of Medinipur. The Subarnarekha river brings these to the sea and the Northeast bound off-shore current transports these over wider tracts. The marine and the terrestrial sediments, before forming soils, support many types of biota. First come the halophyta (salt loving or tolerant). Their debris alter the soil to invite xerophytic (drought resistant) plants. This means that once established, each of these biota causes changes in a specific manner in the sediments to create a more conducive environment for their proliferation. As quantitative additions of plant debris proceed, at a certain stage a qualitative transformation of soil takes place to make room for other types of species to grow there.

In spite of all such changes, the original fabric of the sediments retain their identities for millions of years. For example, beyond the coastal dunes of Midnapur, one finds clayey soils supporting many varieties of vegetation. The clay got deposited here because in a past geological age, the base of the sand dunes was an offshore bar that got attached to the mainland to retain a lagoon like water body to the north of it. This lagoon, due to lack of wave action, sustained a condition for the deposition of fine clay. The original character of this clay has changed to some extent, no doubt. The salt has been leached out to a considerable extent. Humus has been added by the decay of vegetation as a function of ageing. But the stiffness of the soil has persisted. This feature of the environment related to plant-soil interaction is studied by edaphologists. Botanists and pedologists both consider edaphology their functional tool of analysis. If we start looking at the way domesticated vegetation has been introduced within the coastal zone, we cannot fail to understand the nature of naïve appreciation of edaphology by the local residents.

In the off-shore zone also, similar interlinkages of facts studied and collected by the different disciplinarians can be detected. The selection of places for the establishment and later flourishing of the marine biota bear similar parallelism. Alteration of the original state in such locations takes place as a function of time, just as on terra firma, biotic succession takes place as a function of time. The task before the manager of environment is to chart these out within the framework of the dynamic system and sub-systems of ecology. Here, the territorial orientations of the individual domains of sub-systems and system is best achieved through the application of the tools of geomorphology.

A Note of Caution

Before concluding this short article, it is necessary to point out that of all the sub-systems in the coastal zone, the water front is the most sensitive to human impact. Here, the marine and terrestrial materials and nutrients for the growth and survival of the biota get mixed in a very complex manner. Such tracts of extreme sensitivity are generally marked by mangroves. Similarly important are the spawning grounds of zoophytes. Extreme care should be taken to protect these.

4

Russia in Transition: Widening Regional Income Disparities

R.G. Gidadhubli

Introduction

After the disintegration of the U.S.S.R., Russia has been passing through a phase of transition from a socialist politico-economic and social framework to political democracy and a market economy. While the process has been painful and incomplete, it is a matter of great academic interest and practical significance to understand the various issues of the Russian experience of transition. Among the several issues concerning this historic development, one which deserves particular attention is the regional dimension. To understand this problem in its proper perspective, it is worthwhile to briefly recapitulate the regional dimension of economic development during the Soviet period.

Achievements during the Soviet Period

Regional policies and development during the Soviet period were carried out within the framework of centralised planning.

The interests of the regions were to be dovetailed with the interest of the country as a whole through the application of concepts such as regional specialisation, balanced and complex development of regions, backward area development and so on. The declared objectives of the Soviet leadership were to eliminate the wide disparities that existed during the Tsarist period and to bring about equality among the regions. Another specific feature of Soviet socialist regional development, as it was claimed, was the integrated approach, encompassing the economic, social, cultural and political dimensions of Sovietal growth. Hence, planning policies of resource allocation and sectoral development, nationality policy, policies with regard to education, health, etc., were all heavily loaded with regional dimensions and perspectives.

During the 70 odd years of Soviet rule, the leadership of the country could achieve only partial success in its declared objectives and goals. Wide disparities that existed during the Tsarist period between the European and Asiatic parts in the fields of general education, public health and overall economic development were substantially reduced. However, the declared objective of the removal of regional economic differences and bringing about uniformity in income levels among regions was not achieved. In fact, various studies conducted by Soviet and Western scholars indicated that regional disparities widened since the 1970s. Quantitative analysis of regional economic development using indicators, such as per capita income, per capita consumption and per capita investment showed that the Central Asian region, comprising Uzbekistan, Kirghizia, Tajikistan and Turkmenia were 35-40 per cent lower than in the Baltic region, comprising Latvia, Estonia and Lithuania. Similarly, in the case of social indicators, such as number of graduates per 1,000 population, infant mortality per 1,000 births, the Central Asian region was worse off as compared to the European parts. It was claimed, partly justifiably, that complete equality among regions was not feasible due to various objective factors, such as differences in resource endowments, demographic specificities and sociocultural values. However, shortcomings in the implementation of economic policies also contributed to the aggravation of regional disparities. In particular, Mikhail Gorbachev's policies of perestroika and glasnost did not provide for due emphasis on the regional aspects of economic development. On the other hand, as opined by Gertrude Schroeder, Gorbachev's 'misguided policies of democratisation and greater autonomy for the republics fuelled the smouldering fires of latent separation' which ultimately led to the break-up of the Soviet Union.

Post-Soviet Russia

After the Soviet disintegration, 15 former republics have emerged as sovereign, independent countries. Russia, which is the largest among them, has been facing several problems, including the widening regional disparities within its territory. This is evident from the statistical data released by the Goskomstat (State Committee on Statistics) of the Russian government on the per capita income level for 88 administrative divisions (which are also called 'rayons' in Russian) for the years 1993-95 (first half). The summary results are given, in Tables 1 and 2.

Table 1

Number of Regions in Russia by Income Levels

	1993	*1995 (first half)*
Below Average	39	52
Average	24	19
Above Average	25	17
Total	88	88

Table 2
Income Levels in Russia

(in 000 roubles per capita)

	1993	*1994*	*1995 (First Half)*
Average	16.9	49.9	97.9
Lowest	8.7	4.6	10.3
Highest	208.7	250.6	486.4

From the above data, it is clear that the number of poor regions (i.e., below average) has increased significantly, while there has been a modest decline in the number of average and rich regions. This means that in less than two years, the standard of living of Russians in more than half the country has dipped. Similarly, while average income has increased by about six times, the income levels in the poorest region of Russia have hardly increased in about two years. What is worse, the disparity in income levels between the regions with the lowest and highest incomes has widened from 23 times in 1993 to 47 times in 1995. The North Caucasian region has emerged as the richest in per capita income levels.

Causes of Widening Disparity

It may be worthwhile to understand the causes of the widening income disparities in Russia. In post-Soviet Russia, regional variations in income are the reflection of the effect of both the Soviet disintegration, which broke old economic ties, and the politico-economic reform policies initiated by Boris Yeltsin, the president of Russia. Following the prescription of shock therapy to transform the economy, Yeltsin ended overnight all policy instruments of central planning, state control, state ownership, state regulation of prices and initiated the process of decentralisation, with free market prices, and privatisation being the sheet anchor of his reforms. But, the impact of these measures has been uneven over the regions and sectors in Russia.

A few illustrations may be called for. In the past, the Soviet planners had set up huge capacity industrial plants in the Ural, Siberia and Volga regions to produce tractors, automobile tyres, refrigerators etc., to meet the needs of the country; in the process they were getting the advantages of the economies of scale. With the Soviet disintegration, old links were broken, both among the republics and even within Russia among enterprises and, hence, the market for many commodities was affected, forcing them to cut down production which, in turn, affected the employment and income levels of the workers.

Similarly, there are several oblasts and regions, which have only one or two industrial complexes producing defence equipment. In fact, this was a deliberate policy in the past. But with the end of the Cold War, the demand for defence items has declined sharply and these units are finding it difficult to convert to civilian production. Hence, the production in such regions has gone down and even specialised and skilled workers in these units are unable to relocate themselves or find alternative jobs.

The Ivanovo oblast in the Central region in the European part of Russia which specialises in the textile industry, is another interesting case. The units here depend entirely on cotton from Uzbekistan. After the disintegration, Uzbekistan prefers to export cotton to the West for hard currency, so the textile units in the Ivanovo oblast are badly affected, resulting in high unemployment and rendering the oblast one of the poorer regions in Russia.

Regions which have a high concentration of engineering units and factories producing consumer goods are affected considerably because demand for these products has gone down, partly due to competition from imported goods and partly due to the recession in the country. Some of these regions are the Central region, the Volga region and the Central Black Soil region, all in the European part of Russia.

In contrast to the above, there are a few regions that are rich in oil, natural gas, non-ferrous minerals and precious minerals such as gold and diamond, which have the opportunities to export their products; hence, the economic activities in these regions are flourishing and the incomes of the people have risen significantly. Mention may be made of the Yamalo-Nenets region in West Siberia and the Yakutia region in the Far East, which have oil and diamond resources to boost the regions' economic development.

What has added to the disparities is the fact that under the prevailing conditions of decentralisation and federalisation, administrators in some better off oblasts in Siberia and the Far East and even the St Petersburg City administration do not pass on the taxes collected to the Federal budget and often openly defy the Federal government rules. Thus, the worse off oblasts are left in the lurch. Therefore, balancing national and regional interests has become a delicate and difficult task for promoting market reforms in Russia. This was a new problem for Yeltsin as regional income disparities of this magnitude never existed in the Soviet era. Due to the growing disparities, many regions such as the Caucasus, Buryat, etc., in Russia are facing social problems like destitution, poverty, beggary and crime rates of severe proportions. As a consequence, the destitute flee to the large cities for survival by begging. There is also an ethnic dimension to this problem since many of the people living in Chechnya, Dagestan, Tatarstan, etc., are Muslims. Moscow at present is a classic example of this phenomenon. These problems have increased during the last three or four years due to Yeltsin's shock therapy policies, which have shaken the Russian society and their resolution may take several decades.

Remedial Measures

There seems to be some realisation of these social issues on the part of the policy makers, particularly with reference to the 'depressed' and 'crisis' regions of Russia. The Russian government proposed some measures during 1995-97, such as initiating special Federal programmes for socio-economic development, preferential credit and tax concession facilities to enterprises in the crisis and depressed regions, creating additional funds for retraining workers, and so on. But, the declaration of these policy measures appears to be, at best, lip service to tackling the problems facing Russia. At any rate, in the short run, the ability of the Russian government seems very limited in the prevailing tight financial conditions in the country.

5

Revisiting the Growth of Foreign Collaborations in Indian Industries in the Pre-liberalisation Era: A Spatial Analysis of 1970-90

S. Banerjee-Guha

Introduction

Expansion of the economic space to a global scale by international capital is essentially a geographical phenomenon. This expansion works on a continuously restructured absolute and relative space, at both local and global scales. The contemporaneous spread of multinational corporations (MNCs) – one of the most significant organisational forms of international capital – over a much variegated geographical area thus becomes crucial. Multinational corporations create not only an expanded economic environment for capital, but also organise a much divergent use of space and its attributes. The strategy on which the above spatial organisation is designed thoroughly rests on a process of space-specific revaluation and devaluation, which goes to form a part of the internationally operative process of accumulation and underdevelopment (Banerjee-Guha, 1997). It becomes imperative, therefore, to develop an understanding of the strategy of international capital with a spatial perspective, which can focus on the increasing significance of the convergence of economy, society and space as a part of the debate of time-space convergence. With this viewpoint, the present study attempts to examine the spatial growth of foreign collaborations in India – a precursor to present-day globalisation – during the two decades preceding the liberalisation era. The paper is divided into four sections. After introducing the theme in section one, section two examines the regional preference of collaborations in general from five countries (selected for this study). Section three analyses the locational dynamics of foreign collaborations in the major industrial sectors. Section four sums up the observations and relates the foreign corporate strategy with the settlement hierarchy and the prevailing levels of development in India.

The Industrial Policy of India in 1970 intended to augment technological development and exports in the core sector with the help of foreign capital. To achieve this goal, foreign financial participation was permitted even in selected core sectors. Despite the reservations expressed by many (Subramanian, 1972; Bagchi, 1976) about the actual benefits of this relaxation, sectors that had been earlier earmarked for promoting small scale industry, such as leather footwear, toothpaste, household electrical gadgets, paint, etc., were opened up for foreign collaboration. At the same time, this change in policy brought in the Foreign Exchange Regulation Act (FERA) in January, 1974 for a proper monitoring of foreign direct investment (FDI).

Earlier, no proper legal definition of foreign companies existed. By virtue of this Act, the foreign companies in India, except a few selected categories operating in high priority areas or using sophisticated technology or exporting more than 40 per cent of their products, were required to get converted into rupee companies with Indian participation ranging between 20 and 60 per cent, depending on the nature of their activities. Before FERA came into being, the Indian government had delicensed 21 industries in key sectors like drugs, cement, chemicals, fertiliser, paper, etc. Under the suggestion of the Associated Chambers of Commerce and Industry of India (ASSOCHAM), four levels of technology were identified in order to allow foreign participation in various degrees. However, even after the introduction of FERA, a number of giant foreign corporations operating in the key sector of manufacturing, floated separate rupee companies and continued to earn a profit from alternative sources, such as transfer pricing, technology transfer and royalty earnings. The fallacious content of FERA actually promoted foreign exchange remittance by foreign corporations from the mid-1970s onwards, instead of reducing it. Different parent countries, according to their control on the international economy, designed different growth strategies and intensified collaborations (Chaudhuri, 1979).

Between 1976 and 1981, collaboration ventures from U.S.A. increased by 29 per cent, from the U.K. by 17 per cent, West Germany by 27 per cent, France by 32 per cent and Japan by 19 per cent. In the early 1980s, the rate sharpened and all these countries exhibited a rise in their collaboration units. During 1981-86, American collaboration increased by 47 per cent, British by 28 per cent, West German by 48 per cent, French by 36 per cent and Japanese by a huge 71 per cent (Figure 1). By the mid-1980s, American (27.06%) and West German (29.74%) collaborations had almost the same share as the British (29.74%) in the total foreign collaborations in India. Japan had a 12 per cent larger share than it had ever had (Banerjee-Guha, 1998).

Regional Preference of Foreign Collaborations

The participation of the major regions in the collaboration ventures during the 1970s reflected a pattern of pronouncement

Figure 1
Growth of Foreign Collaborations Countrywise (1971-86)

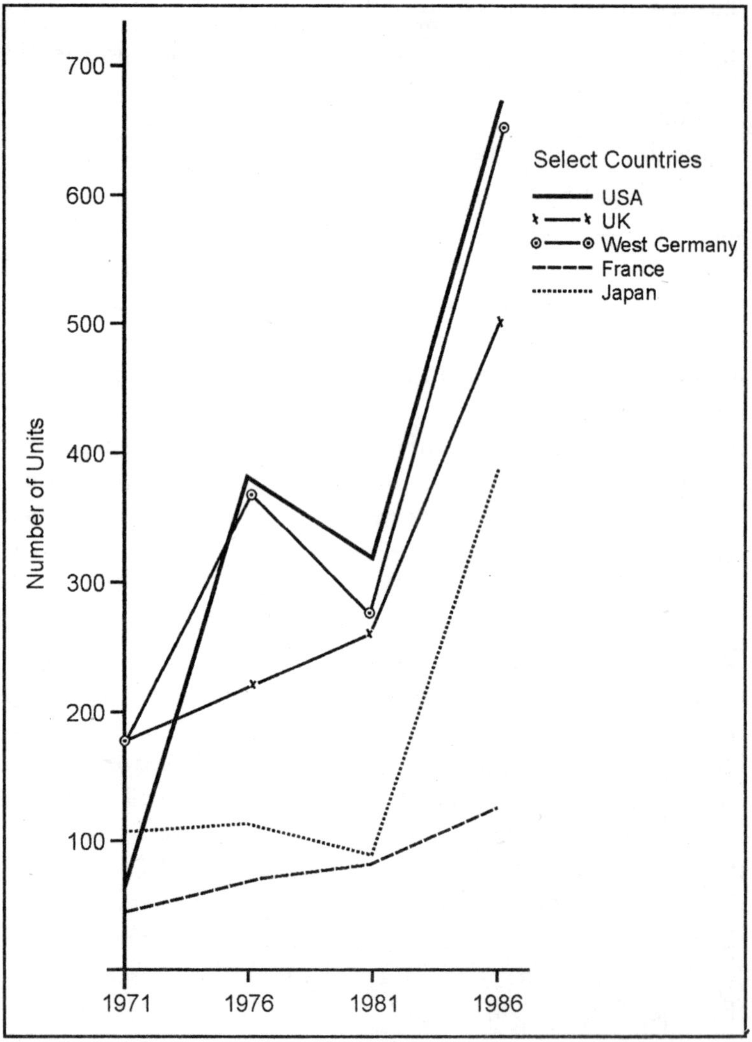

and strengthening. The Western region remained the most favoured region for foreign capital with an average of 48 per cent of the total collaborations during 1976-86. Although the Eastern region still remained the next in importance, it was obvious that it was losing ground to the Northern and Southern regions. In the mid-1970s, the Eastern region accounted for 19.96 per cent of the total foreign collaborations, while the Northern and Southern regions shared 16.58 per cent and 14.17 per cent, respectively. In 1986, the share of the Eastern region remained almost the same (19.43%), but the share of Northern and Southern regions increased to 17.38 per cent and 17.71 per cent respectively.

In the early 1970s, all the five countries had most of their collaborative firms in the Western region. Among them, the heaviest concentration was of the American collaborations with 61.10 per cent of their total collaborations in 1971 located in the Western region. This was followed by West German and Japanese collaborations. Both preferred the Western region with 50.23 per cent and 50.28 per cent of their total collaborations located there (India Investment Centre, 1981).

Although the British and Japanese capital also had a large number of collaborations located in the Western region, they had a substantial share located in the Eastern region, too. West Germany's second priority location was the North and, by 1976, the American collaborations also substantially increased their number (from 7.72% in 1971 to 10.65% in 1976) there. This shift coincided mostly with the reduction of their share primarily in the Western region and also in the Eastern region. In the West, the percentage share of the American collaborations came down substantially from 61.10 per cent in 1971 to 50.90 per cent in 1976. All other countries increased their share of collaborations in the Western region from 1971 to 1976; this got intensified in the 1980s. However, alternate locations for the American, West German and French collaborations were

Figure 2

Penetration of Select Parent Countries in the Four Regions of India

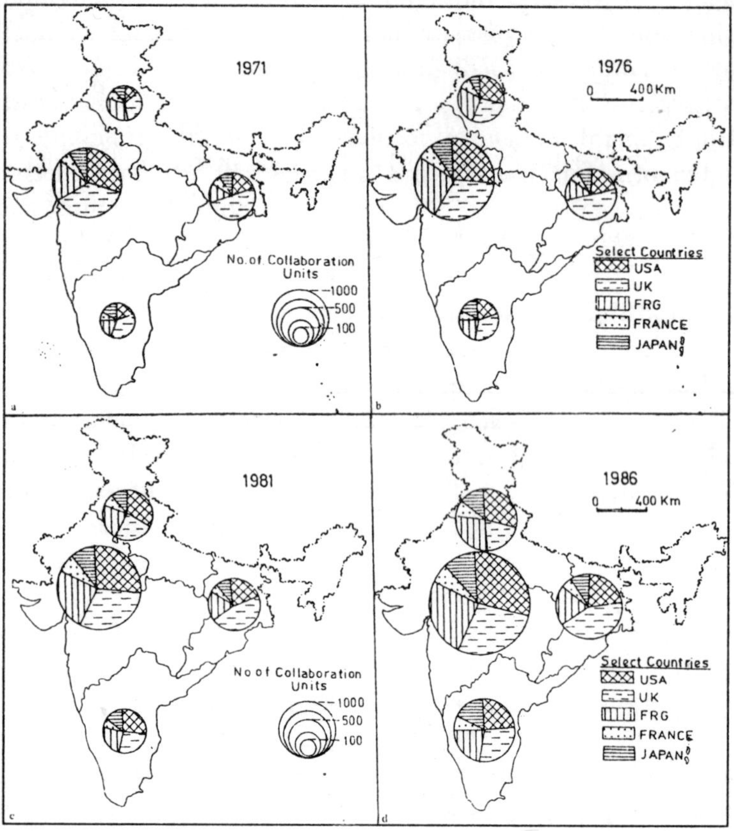

found in the North, for the British in the East and for the Japanese in the South. This pattern got strengthened over the years. Japanese collaborations in the South increased from 19.46 per cent in 1981 to 26.33 per cent in 1986 because of reduction of its investment in the Western and Eastern regions. Since the mid-1980s, West Germany also started preferring the Northern region as a second priority area and increased its investment there. The American and West German collaborations also had similar patterns of locational preference. Almost similar percentages of their total collaborations were located in different regions of India, with the West being the most preferred. The pattern was similar for the French collaborations. The British collaboration firms, however, had a distinctly different locational pattern. For them, the Eastern region was the second priority location followed by the South, both in 1981 and 1986. In the mid-1970s, British capital enjoyed a special status in the Eastern region, accounting for 50.51 per cent of the region's total collaboration undertakings. This concentration pattern later reduced to 43 per cent in 1986. In the Northern, Western and Southern regions, all three countries, the U.S., the U.K. and West Germany, had equal status in sharing regional collaboration ventures, ranging from 21 per cent to 31 per cent during 1976-81 and 22 per cent to 30 per cent during 1981-86. From 1986 onwards, Japan rose to a significant position in the Southern region (Figure 2).

The late 1970s and early 1980s can be identified as a watershed for all these countries in increasing their rate of collaborative ventures. Ironically enough, shortly after the declaration of the 1983 Technology Policy Statement, focusing on the need for research and development and achieving self-reliance, foreign collaborations rose to unprecedented levels (RUPE, 1992). From a little less than one collaboration a day, the rate doubled during the 1980s (Table 1).

Table 1
Foreign Collaborations Approved

Year	*No. of Collaborations Approved*
1960	380
1970	183
1975	271
1981	389
1983	673
1985	1024
1987	853
1989	605
1991	950

Source: RUPE, 1992.

The share of foreign collaborations increased in all four regions. Metro locations got strengthened with varying degrees of intensity. During the 1970s, all the three regions of North, West and East had a concentration of foreign collaboration firms in their regional capitals. However, while Delhi and Mumbai accounted for 82–84 per cent of their respective regional shares in 1971 and 1976, Kolkata had a whopping 95 per cent share of the East. The Southern region experienced a perceptible diffusion in the pattern of location of foreign collaborations by then – a trend observed way back in the 1960s. Chennai accounted for about 52 per cent of the regional share, while Bangalore emerged as a significant alternative. With active infrastructural and governmental support, it became an attractive location for electronic industries with large-scale dependence on foreign technology. In 1976, Bangalore had 22.07 per cent of the regional total. Compared to Bangalore, Ahmedabad and Pune provided weak competition to Mumbai in the Western region. It is interesting to note that among the parent countries, the U.S., West Germany, France and, more specifically, the U.K. had the highest degree of concentration in Delhi and Mumbai in the Northern and Western regions, respectively, while Japanese investment was more mobile. In the Eastern region, however, Japan concentrated almost 95 per cent of its total regional investment in Kolkata. The above pattern of extreme concentration in Kolkata brings out the problem of lopsided growth of infrastructure in the Eastern region (India Investment Centre, 1987).

In the 1980s, with widening opportunities and reducing restrictions, MNCs intensified their metropolitan dominance in locational decisions. In the process, Mumbai emerged as the highest priority location for all the five parent countries. Private capital always enjoyed a secure position in Mumbai. In the past, even in planning endeavours, the Maharashtra government had invited private capital to take active part. Given this background, and with increasing concessional policies, including opening up of the Santacruz Electronic Export Promotion Zone (SEEPZ), foreign collaborations identified Mumbai as their biggest ally. In 1981, Mumbai accounted for 80 per cent of the regional total collaborations and 38 per cent of the entire country's collaborative ventures.

In both the 1970s and 1980s, the British collaborations were significant in Chennai and Kolkata, but their overall share was less than 50 per cent of the total collaborations in these cities. The Japanese collaborations were also steadily increasing their share in the four metropolitan cities, especially in Chennai and Delhi. For Japan, Southern India has always been a favoured location. In the cities of the other regions, Japanese collaborations are few. West German collaborations were significant in Delhi and Mumbai, accounting for 30 per cent and 23 per cent in 1981, and 31 per cent and 23 per cent in 1986, of total foreign collaborations in these two cities.

By the late 1980s, the importance of Mumbai and Delhi as foreign collaboration locations was a recognised fact. In the South, with spread of benefits, the primacy of Chennai as a priority location was considerably reduced. Compared to the other regions, Kolkata and the Eastern region offered a different picture. Despite a flight of capital since the late 1960s and a reduced status in the locational hierarchy, the city was the only foreign collaboration location in the entire region till the mid-1980s. Other urban centres in the East were rarely considered possible locations. Kolkata was actually a poor core of a poor region that was preferred primarily only by British firms in specific sectors.

Locational Dynamics of Foreign Collaborations

The spatial map of foreign collaborations in India in the 1970s brought out the tremendous significance of the Western region. This region emerged as the priority location for twenty-one major and minor industries. It accounted for almost 50 per cent of the collaborations in India in the sectors of chemicals, electronics, machinery, engineering, iron and steel and consumer goods. For pharmaceuticals, the Western region was not just a priority location in the mid-1970s, it was almost the sole location with 90 per cent of the country's pharmaceutical collaboration units located there.

The Eastern region was the second priority region for almost all the industries till the mid-1970s, except electronics, automobiles and consumer goods. Subsequently, due to the militant labour movements and economic uncertainty of this region, the new entrants in these industries preferred other regions. The older industries continued there, primarily out of locational inertia. From the 1980s, however, a major spatial shift was noticed. In the electrical and engineering sectors, the region lost its status as a second priority location to the South and North, respectively. Innumerable engineering units closed down in the Howrah area of Southern West Bengal. Many indigenous firms shifted their head offices and plants from the Eastern region. As mentioned earlier, the 1980s also saw the consolidation of the Western region as the priority location for the major 22 industries of the country, except paper. The pattern brought out the significance of factors like supportive infrastructural base, on the one hand, and social, cultural and political situations, on the other. The active conglomerate of the Indian Industry and Business Class - a product of this region - contributed considerably to its emergence as the centre of corporate capital in India.

The U.S. preferred the Western region in the 1970s for almost all its collaboration ventures in all types of industries (except paper) as did West Germany (except iron and steel, and leather). Britain, France and Japan, however, did not consider the Western region as the only priority location for all their major industries. It needs a mention that West Germany had collaborative associations with the Eastern region in the iron and steel sector since the early planning period. In the automobile and related industries, Britain, France and Japan chose the Southern region as their priority location. France had no collaboration in the electronic sector in the Western region in the mid-1970s. In the 1980s, all five parent countries preferred the Western region for their collaborative ventures in machinery, engineering, iron and steel and pharmaceuticals as well as in rubber, printing and plastic. Exceptions were few: For example, in electronics and automobiles, Britain and France and, especially, Japan set up more collaborations in the South, while for electrical equipment, they favoured the Eastern region. A different pattern emerged for consumer non-food industries. While the American firms preferred the North, West Germany and Britain preferred the South and East, respectively. These shifts essentially indicated a change in the organisation of the economic space. For example, with the emergence of Bangalore as a significant centre in the 1980s, a shift was noticed towards the South from the West. Similarly, with the National Capital Region developing with governmental support, Delhi emerged as an alternate location (Figure 3).

The above pattern also highlights the significance of metropolitan cities, especially the regional capitals, as preferred locations for foreign collaborations in various key and core industries. Except in the South, where Bangalore emerged as an

Figure 3
Major Foreign Collaboration Locations, Regionwise

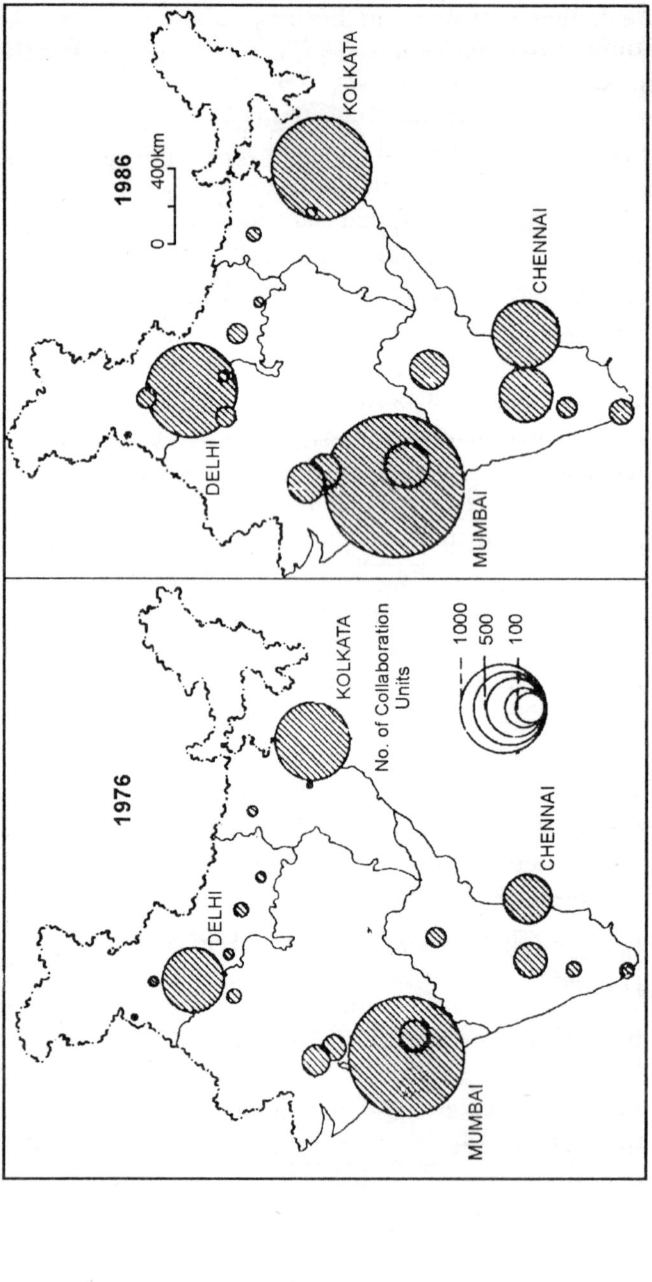

alternative metropolitan location, this trend continued in the 1990s. Pune, Ahmedabad and Surat in the Western region or Rohtak, Chandigarh and Kanpur in the Northern region were not comparable alternatives. Among all the regional capitals, however, Kolkata's supremacy in the Eastern region was the maximum. In almost all sectors, it accounted for 85-90 per cent of the total regional collaborations.

Parent countries that showed a perceptible bias in the mid-1980s towards colonial port locations and regional capitals were Britain and the U.S., especially in key sectors. West Germany and Japan were found to consider alternative locations to Delhi even in a key industry like chemicals in the Northern region. Japan also considered alternative locations in the Northern and Western regions in engineering, iron and steel and machinery. In the Southern region, the only parent country that showed a major bias towards regional capital location was Britain, which can be explained by colonial ties. The general trend for all these countries was not only metropolitan, but the largest city locations. In the South, with the availability of comparable infrastructural facilities in other urban locations, this monocentric trend was considerably reduced.

No major industrial dispersal in interior areas was noticed and no diffusion towards small or medium ranking settlements was identified. Economy of scale played a key role in all the locational decisions, belying the purpose of industrial dispersal or regional development.

Spatial Strategy of International Capital in India during the 1970s and 1980s

A conducive environment is a major requirement of foreign collaborations in any country in order to exploit the economy and space. In India, it was only in the initial planning period that foreign capital faced some resistance. Subsequently and over the years, even in key industrial sectors such as chemicals, engineering, pharmaceuticals and drugs, automobiles, iron and steel and transport related industries, foreign collaborations expanded in a steady manner.

Analysis of the locational pattern of foreign collaborations in India brings out the dynamics of accumulation in a country, rather than of static allocation of activities to their best locations with respect to market, labour and material (Walker and Storper, 1981). During the 1960s, besides Mumbai's emergence as an industrial capital through successive stages of production and investment, there was also a spatial shift of industrial concentration from the East to the West. This was certainly an additive factor for the rise of the Western region, especially Maharashtra, as a major industrial region of post-colonial India. In 1961, the Western region already accounted for more than 30 per cent of the total registered factories of the country, with Maharashtra alone accounting for 17 per cent. During this period, the Eastern region, too, accounted for nearly 24 per cent of the total factories. In 1971, the relative position of the Western region remained almost the same, with 31 per cent of the total registered factories of the country, while the Eastern region's share came down. By that time, the Southern region, with Bangalore as an emerging industrial centre, offered an alternative to the Western region in both small and large industries, accounting for 31 per cent of the factories. Although during the 1980s, the North experienced a substantial rise in the number of factories, yet in terms of total number of factories, the most significant feature of the period was the emerging competitiveness of the South vis-à-vis the West and the further decline of the East. In 1986, the Eastern region accounted for only 14 per cent of the total registered factories of the country, while the Western and Southern regions accounted for 33 per cent each. The statewise distribution shows that in 1986, Maharashtra alone accounted for 13 per cent and Gujarat 9 per cent of the country's total, while Karnataka, Andhra Pradesh and Tamil Nadu accounted for 8 per cent, 10 per cent and 8 per cent, respectively. This also shows that from the 1970s, Maharashtra's share declined (from

17% in 1971 to 13% in 1986), while those of Tamil Nadu, Gujarat and Karnataka increased.

The spatial distribution of foreign collaborations also followed a similar pattern of factory location. It showed a bias towards the Western states, especially Maharashtra and Gujarat, with an increasing shift towards Karnataka, Andhra Pradesh and Tamil Nadu. The second priority was West Bengal in the East and Delhi in the North. The focus was on selected states and selected locations, with little concern for resource location or levels of development, but much to do with the system of production, investment and accumulation. The percentage rate of increase of foreign collaborations from 1971 to 1986 was also much higher (101%) than that of total factories (85%). Most of the favoured states, like Maharashtra, Karnataka, Gujarat, Tamil Nadu and West Bengal, showed positive deviations from the national percentage change of foreign collaborations from the mid-1970s to the late 1980s. Similar deviation studies undertaken by Dicken and Lloyd (1980) on foreign controlled manufacturing employment in the U.K. between 1963 and 1975 show a substantial narrowing of regional differentials and the relative decline of some regions in the Southeast.

The phenomenal geographical concentration of foreign collaboration units in a few locations also brings out the significance of labour factor heterogeneity in location decisions. In the early 1980s, Storper and Walker (1981) worked on the heterogeneity of labour and distinguished the several dimensions of its variability as (i) conditions of purchase, including wage level, work, work rules etc., (ii) performance capacity, including technical skills and discipline as that affects job performance and creativity, (iii) labour control and actual performance, supervision or monitoring to counter workers' resistance, and (iv) reproduction or generation of identifiable work characteristics that can persist over a time period. In India, while labour can certainly be considered as a factor of location and spatial shift, it is equally true that the political economy of labour (Malecki, 1985), including its complex network of heterogeneity, works more than the singular factor of labour skills and resources. A salient example is the Eastern region, which has not been chosen by foreign collaborations as a location in the later years of the 1980s, even in sectors like iron and steel or engineering, which have a long tradition of growth in this region due to the proximity of resources and labour skills. Natural resources have not figured as a major factor for these firms as compared to the factors of efficient processing opportunities, available substitutes, expansion of assembly type production and availability of material inputs in the form of produced commodities available in other regions. These plants thus became footloose and enjoyed a locational freedom primarily on the basis of inter-industry linkages and labour heterogeneity. In addition, as foreign collaboration firms essentially rank the potentials of communication and information diffusion as other significant locational factors, their metropolitan bias was distinct. In addition, the political situation of the respective states, tariff and non-tariff barriers, tax rates and special investment incentives (Chang, 1989) worked to make these collaborations locate themselves in Western Maharashtra, Southern Gujarat and, subsequently, in Southern Karnataka, different parts of Tamil Nadu in the South and in the National Capital Region. The most favoured region since 1971, however, has been the Western region. In spite of the relative locational changes since the 1980s favouring the South, Mumbai and the West continued to remain a prime location (Figure 4) even in the 1980s. The Mumbai-Ahmedabad and Mumbai-Pune regions, which form an elongated belt in the coastal parts of Maharashtra and Gujarat, were identified as the most preferred investment locations for foreign collaborations by almost all the five parent countries till the late 1980s. A parallel can be drawn with the locational significance of the New York Metropolitan Area in the U.S. During the late 1970s, it was the most preferred investment area for foreign firms in the United States (Bagchi-Sen and Wheeler, 1989).

Figure 4
Metropolitan Bias in Foreign Collaboration Locations in India, 1986

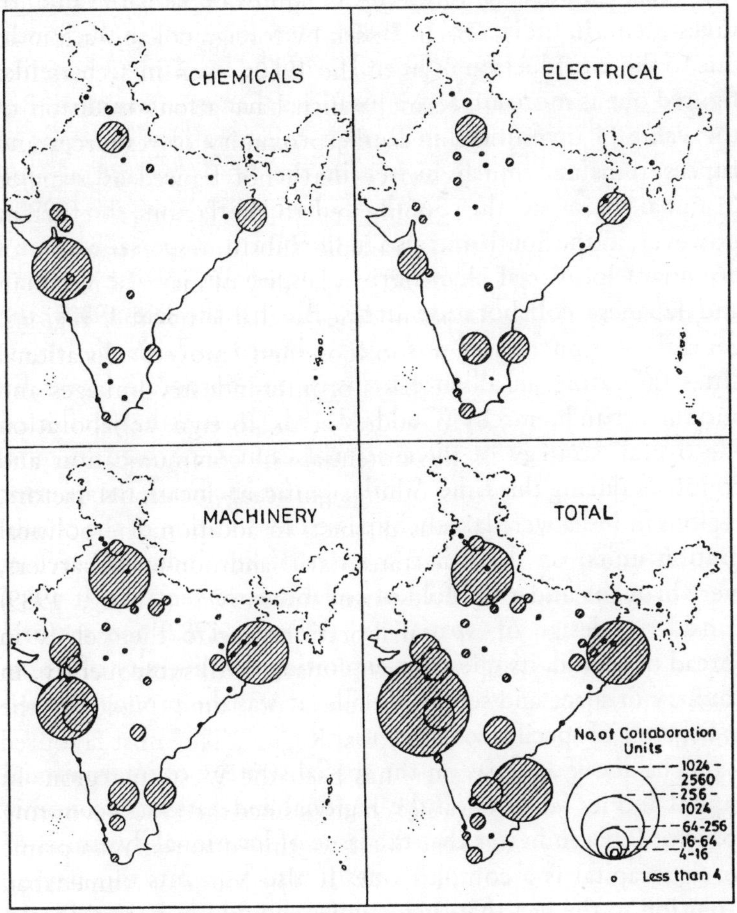

Later, in the early 1980s, the Miami area accounted for a greater number of foreign transactions, although the former remained the most important location in terms of dollar value of investment. In India, too, in 1980s, a few locations in the South, due to the introduction of liberal tariff and investment benefits, figured out as more attractive locations than even Mumbai. But the value of investment in foreign collaborations in terms of rupees remained much higher in the West, in and around Mumbai than in the South. Subsequently, in the 1990s, however, in the South and also in the North, dispersal became a dominant locational characteristic, especially for the German and Japanese collaboration firms. But till the late 1980s, the general locational trend was overtly biased towards the largest cities indicating an alliance between the corporate hierarchy and the urban hierarchy (Goddard, 1977). It strikingly reflected the overall strategy of international capital, pursued in most countries during this time. While, on the one hand, the interior regions in India were largely bypassed for location of the collaboration units, on the other, a series of key industrial sectors were brought under the fold of collaborative ventures. It was a considered design of 'spatial fix' (Harvey, 1982) and sectoral spread of capital, its quest for exploiting the diversity of a host country in space and sector. Finally, it was the product as well as begetter of specific social relations.

The above analysis on the spatial strategy of international capital and its fall-outs on the regional and national economic space of India indicates that the issue of locational dynamics of foreign capital is a complex one. It also warrants our paying attention to the fact that such studies should not be considered an isolated issue of development or economic geography, but one that leans heavily on society-space convergence and, hence, to be studied from an interdisciplinary perspective.

Acknowledgement

The author wishes to thank Prasad Gogate, cartographer, Department of Geography, University of Mumbai, for preparing the maps used in this article.

Note

During the period of the analysis, Germany was still divided. In the above article, only West German firms have been studied.

References

Bagchi, A.K. (1976): 'Deindustrialisation in India', *Journal of Development Studies*, 12(2), 135-64.

Bagchi-Sen, S., and Wheeler, J.O. (1989): 'A Spatial and Temporal Model of Foreign Direct Investment in the United States', *Economic Geography*, 65, 113-29.

Banerjee-Guha, S. (1997): *Spatial Dynamics of International Capital: A Study of Multinational Corporations in India*, Orient Longman, Hyderabad.

Banerjee-Guha, S. (1998): 'Foreign Collaborations in Indian Manufacturing during Early Years of Planning: A Spatio-temporal Analysis', *Geographical Review of India*, 60 (4), 457-79.

Chang, Kang-tsung (1989): 'Japan's Direct Manufacturing Investment in the United States', *Professional Geographer*, 41(3), 314-29.

Chaudhuri, S. (1979): 'Financing Growth of Transnational Corporations in India, 1956-75', *Economic and Political Weekly*, August 18, 1431-35.

Dicken, P., and Lloyd, P.E. (1980): 'Patterns and Processes of Change in the Spatial Distribution of Foreign Controlled Manufacturing Employment in the United Kingdom, 1963-1975', *Environment and Planning*, A-12, 1405-26.

Goddard, J.B. (1977): 'City and Regional Systems', *Progress in Human Geography*, 1(20), 296-303.

Harvey, D. (1982): *Limits to Capital*, Basil Blackwell, U.K.

India Investment Centre (1981): *Directories of Foreign Collaborations*, IIC, New Delhi.

Malecki, E.J. (1985): 'Industrial Location and Corporate Organisation in High Technology Industries', *Economic Geography*, 1(4), 345-69.

RUPE (1992): 'Aspects on Indian Economy', No. 5, 39-55.

Storper, M., and Walker, R. (1981): 'The Spatial Division of Labour' in L. Sawers and W.K. Tabb (eds), *Sunbelt-Snowbelt Urban Development and Regional Restructuring*, O.U.P., U.K., 143-56.

Subramanian, K.K. (1972): *Import of Capital and Technology*, People's Publishing House, New Delhi.

Walker, R., and Storper, M. (1981): 'Capital and Industrial Location', *Progress in Human Geography*, 5(4), 473-509.

6

Urban Environment, Sanitation and Health in India

Jayasree De

Introduction

India's urban population of over 217 million, constituting nearly 26 per cent of the total population of the country, is one of the largest in the world. The trends of urbanisation, in terms of increase in both urban areas and population, show a steady progress from the beginning of this century (Table 1). This population, therefore, constitutes an important target group whose well-being will determine the developmental process of the country. However, the phenomenal increase in urbanisation in India has imposed a considerable strain on the life support system in cities. This has created problems of environment, which, in turn, have had an adverse impact on the health and well-being of the population residing therein.

Health, as is well known, does not imply mere absence of disease. It symbolises harmony between man and his environment (Howe, et al., 1986), which results in a state of complete physical, mental and social well-being. It is, therefore, necessary to understand the conditions of the urban environment in India in order to comprehend the state of sanitation and health prevailing in these areas.

Table 1
Trend of Urbanisation in India, 1901-91

Census Year	*No. of Urban Agglomerations*	*Urban Population (in million)*	*Urban Population (%)*	*Decennial Growth (%)*	*Annual Exponential Growth (%)*	*Annual Gain in Urban Population (%)*
1901	1,827	25.25	10.84	-	-	-
1911	1,815	25.94	10.29	0.35	0.03	-0.06
1921	1,949	28.09	11.18	8.27	0.79	0.09
1931	2,072	33.45	11.99	19.12	1.75	0.08
1941	2,250	44.15	13.86	31.97	2.77	0.19
1951	2,843	62.44	17.29	41.42	3.47	0.34
1961	2,365	78.94	17.97	26.41	2.34	0.07
1971	2,590	109.11	19.91	38.23	3.21	0.19
1981	3,362	159.46	23.34	46.14	3.83	0.34
1991	3,768	217.18	25.72	36.19	3.09	0.24

Source: India Economic Information Year Book 1991-92, p. 77.

Urban Environment

Environment refers to the totality of external influences – natural and man-made – which impinge on man and affect his well-being. It thus embraces the life support systems and also the multiplicity of stimuli and hazards, direct and indirect, which man experiences (Howe, 1976). The urban environment is an interaction between physical, biological and sociocultural elements that are in a dynamic process of transformation. The phenomenon of urbanisation is man-made, being only indirectly influenced by natural conditions. Hence, the characteristics of the urban environment are largely human creations that fashion the life support system and bring about direct and indirect hazards that pose a challenge to man's health and well-being (Giggs, 1979; Eyles, 1982).

Urban areas in India are characterised by a high concentration of economic activities, high density of population, unabated immigration, mushrooming of squatter colonies and increasing scarcity of basic amenities and infrastructure. The population of these areas comprises predominantly people who migrate to the city from the countryside either in the hope of better economic opportunities or in the expectation of a better lifestyle. In India, there is a clearly discernible dichotomy between the lifestyles of the rural areas and the more modernised living conditions in the cities. Adjustment to urban living is thus beyond the capacity of the migrant population that gets drawn into it, resulting in unhygienic living conditions and insanitary habits.

With population increasing in the cities, there is a chronic shortage of housing. Since the majority of the population moving into cities belongs to the economically deprived strata, and since there is no provision to accommodate them, this population is compelled to live in improvised shacks, resulting in the growth of slums. These unauthorised hutments usually spring up in the most uninhabitable sites in cities, such as along river channels or 'nallahs', or on unused land that cannot be put to any productive use. They also occupy open spaces, thereby nullifying the basic function of such open grounds. Most of these colonies, devoid of water supply, drainage and other basic amenities, are veritable cesspools of disease.

According to the Census of slums conducted in 1981, the slum dwellers in urban areas account for an average of nearly 19 per cent of the total urban population (Table 2). However, the break-up of the slum population, according to city size classes, reveals that there is a direct relationship between population size class of cities and the percentage of slum population. Urban centres with a population exceeding one million have an average of about 31 per cent slum dwellers, although many growing cities, such as Vadodara are known to have as high as

50 per cent of its population living in slums. The Eighth Five Year Plan indicates that there were 48.8 million people living in slums in 1990, 40 per cent of which were in million plus cities. The population at risk to diseases and ill-health is thus very high in large cities.

Table 2

Estimates of Slum Population by Size-Class of Urban Agglomerations/Cities/ Towns in India, 1981

Population Size Class of UAs/Cities/ Towns	*No. of UAs/Cities/ Towns*	*Urban Population ('000)*	*Estimated Slum Pop. ('000)*	*Slum Population (%)*
1 million and above	12	42,024	12,934	30.08
500,000 – 10,00,000	29	19,364	3,985	20.06
300,000 – 500,000	30	10,905	1,935	17.7
100,000 – 300,000	150	23,462	4,252	18.1
50,000 – 100,000	275	18,550	2,256	12.2
Below 50,000	2,866	45,088	4,527	10.0
Total (All Classes)	3,362	1,59,393	29,889	18.8

Source: India Economic Information Year Book, 1991-92 p. 79.

Shortage of housing is a problem that affects all strata of population in cities. The haphazard and unrestricted growth of housing colonies in the peripheral areas of cities results in not only frequent expansion of city limits, but also the disappearance of the green belt (De et al., 1987). Owing to the pressure of population on land, housing design and layout have dispensed with open spaces around houses, converting the landscape into a mass of concrete. Most of the new housing colonies spring up even before roads, electricity, drinking water supply, drainage and other infrastructure can develop. This leads to illegal tapping of existing resources from the nearest available point. Such temporary connections do not conform to standard specifications. So it is not uncommon to find them collapsing frequently. Epidemics of gastrointestinal diseases resulting from contamination of drinking water with sewage are, therefore, becoming increasingly common in Indian cities. Limitation of space is the main physical condition resulting from overcrowding. This is noticeable, not only from the density of houses, but also from the crowding of houses. Studies conducted in various cities have revealed that social and physical ills in urban centres are attributable to high population density and overcrowding.

Another affect of increasing population in cities and associated unhygienic habits is the increasing accumulation of wastes. The waste disposal system in towns and cities is grossly inadequate, as a consequence of which huge dumps of garbage on the roadsides, even in upper class areas, are a regular sight in almost all cities.

In India, industries tend to concentrate in urban areas for the obvious economic and infrastructural advantages provided at such locations. Unrestricted industrial growth has conferred the added problems of air and water pollution caused by the disposal of industrial wastes. The laws and by-laws for enforcing strict effluent treatment regulations at industrial plants are yet to be made effective. The result is the spewing of chemicals from factory chimneys or disposal of the same through available water channels. According to Rajagopal (1987), out of the 1,700 main polluting industries in the country in the early 1980s, only 357 units had incorporated technology to minimise pollution. Only 7 per cent of industries were found to have equipment to control air pollution and, in Delhi, the air in the Badarpur Thermal Power Plant area was found to contain particulate emissions several times the permissible limits.

Noise from factories is another source of pollution. Although in some cities, large industries have initially been located on the periphery or beyond the city limits, the advancing city has now engulfed the intervening green belt so that housing colonies and industries today are juxtaposed. It is in such situations that industrial pollution can create havoc on the scale that was witnessed in Bhopal.

Industrialisation has brought in its wake increased commercial activities. This, together with growth in population, has resulted in enhanced transport-related activities and increase in the number of trucks and motor vehicles plying in the cities. The number of existing roads do not match the ever-increasing fleet of motor vehicles. The results are traffic congestion, exacerbation of air and noise pollution and a larger number of accidents.

The urban climate is modified by the physical and cultural environment of cities. Concrete structures, paved surfaces, population density and traffic congestion are responsible for modifying the micro-climate in cities through their effect on local thermal and humidity conditions (Subrahmanyam et al., 1982). These micro-level variations in climate, resulting in the formation of heat islands, are peculiar only to urban areas and responsible for irritation to the human body and mind.

Urban Sanitation

In the light of the prevailing environmental situation in urban areas, the condition of urban sanitation is not difficult to visualise. According to the Planning Commission reports, at least 52 per cent of the urban population has no access to proper sanitation and about 15 per cent are yet to be provided with safe drinking water (Table 3). Besides, even those who are served by water connections, obtain their supplies only for short periods in a day. Absence of continuous flow of water enhances the unhygienic conditions.

Absence of a proper sewerage system in the newly developed and developing housing colonies, as already stated, is one of the major problems of sanitation. It not only leads to deterioration of the quality of areas, but is also a root cause of numerous infectious and parasitic diseases. Lack of paved roads in many of these residential colonies leads to accumulation of water and slush during the rainy season, providing ideal

Table 3

Achievement in the Provision of Urban Water Supply and Sanitation at the End of the Seventh Plan (1985-90) and Two Annual Plans (1990-92) (population in millions)

Sub-sector	*Population Covered as on 31.3.85*	*Urban Pop. (%)*	*Population Covered as on 31.3.90*	*Urban Pop. (%)*	*Anticipated Population Coverage as on 31.3.92*	*Urban Population (%)*
Urban Water Supply	127.20	72.90	182.00	83.80	185.67	84.90
Urban Sanitation	49.60	28.40	99.70	45.93	104.76	47.90

Source: Eighth Five Year Plan, 1992-97 Vol. II, Planning Commission, Government of India, New Delhi, p. 378

conditions for the breeding of mosquitoes. These puddles of water and heaps of garbage along the roads facilitate the proliferation of disease inducing organisms, such as bacteria and protozoa, vectors including mosquitoes, houseflies, various orthropods and helminthes. The drainage channels in cities are usually clogged with wastes and industrial effluents, and reduced to sluggishly flowing cesspools, which pose immense health hazards. Stray animals, including cattle, pigs and dogs, which form a part and parcel of cities, are another source of insanitary conditions, resulting in further health risks.

Urban Health

Although official statistics in India are woefully inadequate to show the relative conditions of health in urban and rural areas, independent studies carried out by researchers clearly indicate that urban health conditions are far more diverse and complex than those in the rural areas.

The health status of a population is known to differ in accordance with lifestyle and socioeconomic conditions. Socioeconomic conditions are probably a key factor in health differentials. India is one of the countries that is passing through a phase of transition from an agrarian culture to an industrial society. This has resulted in the plural character of its health and disease patterns. On the one hand, infectious diseases still take a heavy toll of life and account for a large incidence of morbidity; on the other, chronic and degenerative diseases are an ever-growing phenomenon, particularly in urban societies where increasing modernisation and Westernisation are creating more affluent lifestyles, which are in sharp contrast to those of the rural areas.

In urban areas, some indicators, such as infant mortality rates and crude death rates reflect a better quality of life than that in rural areas. However, the reality is masked by considerable intra-urban variations. Actually, disease and ill health are probably more rampant in urban centres. Studies of morbidity patterns in several urban areas have shown the preponderance of gastrointestinal and respiratory problems which are closely associated with the prevailing conditions of environment and sanitation. Similarly, mortality trends in the urban areas of India grouped on the basis of cause of death indicate the dominance of infectious and parasitic diseases as well as diseases of the circulatory system (Table 4).

Viral hepatitis, cholera and enteric fever are the common gastrointestinal diseases. Viral hepatitis is a major public health problem in most cities of India and annual outbreaks of epidemics are a regular phenomenon. Insanitary living conditions and overcrowding, together with poor socioeconomic status, appear to be the familiar conditions associated with the endemicity of this disease. In Vadodara city, for example, 97 per cent of the sufferers were traced to poor quality environments, and 93 per cent belonged to the lower or lower middle classes of society. Around 57 per cent of the cases were found to be living in single room tenements (De et al., 1986). Epidemics of viral hepatitis, almost invariably, are caused by contamination of drinking water with underground sewage. Another point of interest is that epidemics appear to be more common in the

Table 4
Medically Certified Deaths by Major Groups to Total Deaths in Urban Areas of India, 1987

Cause of Death	*Total Deaths (%)*
1. Infectious and Parasitic Diseases	18.85
2. Neoplasms	3.55
3. Endocrine, Nutritional and Metabolic Diseases and Immunity Disorders	2.60
4. Diseases of Blood and Blood Forming Organs	2.30
5. Mental Disorders	0.15
6. Diseases of the Nervous System & Sense Organs	3.75
7. Diseases of the Circulatory System	19.70
8. Diseases of the Respiratory System	9.60
9. Diseases of the Digestive System	4.05
10. Diseases of the Genito-urinary System	1.65
11. Complications of Pregnancy, Childbirth & Puerperium	3.00
12. Diseases of the Skin & Subcutaneous Tissue	0.10
13. Diseases of the Musculo-skeletal System & Connective Tissue	0.10
14. Congenital Anomalies	0.75
15. Certain Conditions Originating in the Perinatal Period	9.45
16. Symptoms, Signs and Ill-defined Conditions	11.40
17. Injury and Poisoning	10.40

Source: Mortality Statistics of Cause of Death, 1987. pp. 12-14. Average of data for 12 states and two union territories.

newly developed housing colonies, where underground sewerage lines are yet to be laid. The relationship between the incidence of this disease and unsewered areas, greater number of borewells prone to contamination as well as slum population, has been established in studies conducted in the cities of Chennai and Madurai (Kumaran et al., 1986).

Cholera is another gastrointestinal disease that manifests in the form of an epidemic, particularly in the summer and post-monsoon seasons (Park and Park, 1991). Contamination of food and water are the main causes. The environmental associations of the disease are similar to those of viral hepatitis. Enteric fever has a similar aetiology. Its association with the monsoon climate, crowded and unhygienic living conditions, water pollution and specific social areas and occupational classes, has been noted in Brihanmumbai (Koshy et al., 1980).

The most prominent respiratory diseases in the urban areas of India are pulmonary tuberculosis, influenza and predominantly infantile diseases such as diphtheria, chicken pox and mumps. Although the spread of these diseases is mainly associated with overcrowding and close contacts, the ecological association of tuberculosis is more complex. Tuberculosis is also one of the major causes of mortality in both the rural and urban areas in India. However, the urban mortality rates are much higher as studies, such as those conducted in West Bengal, have shown (Hazra, 1986). The disease prevalence is high among industrial workers and slum dwellers. Its incidence in urban areas shows a close association with overcrowding in rooms, socioeconomic conditions and atmospheric pollution. Research carried out in Vadodara city has shown that 50 per cent of the cases of pulmonary tuberculosis occur among industrial workers (De et al., 1987). No case is reported from the upper socioeconomic strata, with 94 per cent of the patients belonging to the lower income groups. The majority of the sufferers were found to live in single room accommodation; spatially, there was also a positive correlation between population density and number of reported cases.

Atmospheric pollution is one of the principal causes of the increasing incidence of respiratory morbidity in urban areas. Industries and traffic are the two main sources of atmospheric pollution. A positive relationship between suspended particulate matter and sulphur dioxide in the atmosphere and decreased pulmonary function has been established (Durvasula et al., 1985). Increasing cases of bronchial disease and asthma are attributed to this factor. Pollution of the atmosphere with lead, resulting from increasing traffic density, enhances susceptibility to lead poisoning, particularly among the child population (Patil et al., 1988). WHO studies quoted by Rajagopal (1987) have shown that incidence of respiratory diseases is higher in zones of heavy traffic, where the lead from automobiles is very high, such as in Mumbai and Kolkata. In Kolkata, it is 17 micrograms per cublic metre as compared to 4.8 in London and 7.5 in New York. In Delhi, the number of vehicles is very high and, in the Connaught Place area, the air carries three times the permitted levels of toxic carbon monoxide. In Mumbai, people living in polluted areas suffer from higher incidence of bronchitis, tuberculosis, skin diseases and breathlessness.

Other common urban diseases are orthopod-borne infections like malaria and filariasis, which are exerting immense pressure on public health efforts (Prasad, 1985). The urban environment offers a wide range of conditions suitable for the proliferation of vectors and parasites as well as for the transmission of disease causing organisms. In the case of both malaria and filariasis, the mosquito prefers fresh water for breeding while the mosquitoes transmitting filariasis proliferate in polluted water. Both types of vector habitats are found in the urban areas of India, thanks to poor sanitary conditions, open drains, septic tanks and accumulation of rain water in low-lying areas, cisterns and numerous other breeding places. Therefore, the spatial distribution of the prevalence of malaria is associated with the environmental features conducive to the breeding of mosquitoes as illustrated in a study conducted for Chennai city (Hyma et al., 1983).

In industrial cities, occupational diseases such as silicosis, byssinosis and bagassosis are additional health hazards. Besides, loss of hearing resulting from noise pollution in factories, eye problems caused by exposure to various industrial processes, skin diseases owing to contact with chemicals, stress conditions resulting from tensions at the workplace are some of the common ailments in the industrial cities of India. Owing to the absence of recorded statistics on the prevalence of occupational diseases in the country, the enormity of this problem is yet to receive adequate attention.

According to present trends, non-communicable diseases, such as cardiovascular ailments and cancers are on the increase in urban areas. The affluent lifestyles in cities are largely associated with many of these diseases. For instance, most city dwellers use vehicular transport, thus dispensing with the exercise available from walking. Greater opulence in cities gives rise to easier access to amenities and conveniences which produce more luxurious and sedate lifestyles and impair health in the long run. The increasing percentage of deaths from ailments of the circulatory system and from neoplasm is a pointer to this fact.

Apart from various communicable and degenerative diseases, urban areas in India are further confronted with health situations related to social problems. In most cities, there is a large proportion of workforce which comprises men who have moved to the cities in search of jobs, leaving their families in the villages. This situation has given rise to alcoholism, prostitution and other social evils, with a corresponding rise in sexually transmitted diseases and HIV/AIDS.

The stress and strain of urban living, overcrowding and economic deprivation of a large portion of the population leads to mental and psychological disorders as well as social evils like delinquency, drug addiction and crime. These conditions are closely associated with urban life and jeopardise the social well-being of large sections of the urban community and several localities within cities. These social evils, though universal, are definitely more pronounced in the urban milieu than in the rural environment.

Conclusion

Ill-health in cities is associated with a variety of conditions related to the sociocultural traits of urban areas, including insanitary conditions, pollution from industries, traffic congestion, accumulation of wastes, overcrowding, migration, occupational hazards and the stress and strain of urban living. Hence, there is a considerable intra-urban variation in health conditions determined by the quality of urban areas.

In the ultimate analysis, it becomes clear that human intervention in the ecosystem represented by an urban area, has been responsible for the complex and multiple health problems that characterise Indian cities. The health situation and disease patterns in Indian cities today bear a close resemblance to the conditions that prevailed in European cities in the years immediately following the Industrial Revolution, when men began to move into cities from the rural areas. It was the timely introduction of public health programmes that changed the entire health scenario in those countries. The degradation of the urban environment in India has created insanitary and unhealthy living conditions that have imposed undue stress on the physical and mental well-being of the population. Restoration of health in urban areas would involve a multi-faceted approach to physical and human resource development that can only be achieved through an integration of economic and social welfare schemes involving basic urban infrastructural improvement. The public health programme in India clearly needs to be more broad based.

References

De, J., and Gollerkeri, R.S. (1986): 'Socio-cultural Aspects of Infectious Hepatitis in Vadodara, India', *Asian Geographer*, 5(1), 99-105.

De, J., and Gollerkeri, R.S. (1986): 'Pulmonary Tuberculosis and Its Socio-cultural Setting in Vadodara', *Geographic Medica*, 17, 161-76.

De, J., Gollerkeri, R.S., and Srivastava, A.K. (1987): 'Some Preliminary Consideration of Environmental Planning for Urban Areas' in P.K. Saha (ed.), *Impact of Development on Environment*, Vol. 2, Geographical Society of India, 123-9.

Durvasula, R.S., Shah, D.N., and Rajagopal, T. (1985): *An Epidemiological Study of Air Pollution in Baroda City with Respect to Respiratory* morbidity, Proceedings of Seminar on Hazards in Chemical Industry, Indian Institute of Chemical Engineers, RSD 1-9.

Eyles, J. (1982): 'Health and Medicine in Urban Society: The Social Construction of Fetishism of Health, Contemporary Perspective on Health and Healthcare', Health Research Group, Occasional Paper No. 20, Dept of Geography, Queen Mary College, University of London.

Giggs, J.A. (1979): 'Human Health Problems in Urban Areas' in D.T. Herbert (ed.), *Social Problems and the City: Geographical Perspectives*, Oxford University Press, London.

Hazra, J. (1986): 'Tuberculosis in West Bengal', *Geographia Medica*, 16, 100-21.

Howe, G.M. (1976): 'Environmental Factors in Disease' in J. Linihan and W.E. Fletcher (eds), *Health and the Environment*, Blackie, Glasgow and London, 1-29.

Howe, G.M., and Loraine, J.A. (eds)(1986): *Environmental Medicine*, Heinemann, London.

Hyma, B., Ramesh, A., and Chakrapani, K.P. (1983): 'Urban Malaria Control Situation and Environmental Issues, Madras City, India', *Ecology and Disease*, 2(4), 321-35.

Koshy, M., and Arunachalam, B. (1980): 'Spatial and Temporal Variations in the Incidence of Enteric Fever in Greater Bombay', Paper presented at the 1st Indian Geography Congress, NAGI, Chandigarh.

Kumaran, T.V., and Ananthan, P.H. (1986): 'Infectious Hepatitis in Urban India', *Geographia Medica*, 16, 66-82.

Park, J.E. and Park, K. (1991): *Textbook of Preventive and Social Medicine*, Banarasidas Bhannot, Jabalpur, India.

Patil, P.N., and Gadgil, A. (1988): 'Environmental Lead Distribution in Relation to Road Dust Samples In and Around Pune', *Transactions*, IIG, 10(2), 65-7.

Prasad, S. (1985): 'The Incidence of Malaria in Major Urban Centres of Andhra Pradesh 1975-1984', *Annals*, NAGI, 5(2), 37-43.

Rajagopal, P.R. (1987): 'Environmental Factors in Human Health' in R. Kumar (ed.), *Environmental Pollution and Health Hazards in India*, Ashish Publishing House, New Delhi, 260-2.

Subrahmanyam, V.P., Sachi Devi, S. and Vishwanatham, P. (1982): 'Some Aspects of Urban Climatology of Vishakhapatanam', *Transactions*, IIG, 4(2), 133-41.

7

Urbanisation and Its Socioeconomic Correlates in Maharashtra, 1991

Dipti Mukherji and V.S. Phadke

Introduction

Economic development is accompanied by a shift in labour allocation from the primary to the secondary and, eventually, to the tertiary sector (Clark, 1940). This is amply demonstrated by the experience of the developed countries (Fuchs, 1968; Bell, 1973; Kellerman, 1985). Improvement in technology enhances productivity, making a part of labour in the primary sector redundant, thereby enabling a shift to the secondary and tertiary sectors. Labour productivity also improves in these sectors, but it is slow in the latter as such activities are infrastructure oriented (Gong, 2002). Further, the services to be provided become increasingly sophisticated and they continuously open up new avenues. Tertiary employment thus shows the slowest decline among the three sectors. It also possesses the highest income elasticity of demand and, with advance of the economy, the sectoral transition shows a higher proportion of the labour force in the tertiary sector.

The experience of the developing countries is slightly different from the above. The tertiary labour force in these countries was larger and grew more rapidly than that in secondary activities even in the early stages of development (Pandit and Casseti, 1989). This is often attributed to their late start, enabling them to adopt improved and more capital intensive technology; the growth of the informal sector also contributes substantially to this phenomenon. The small volume and less sophisticated nature of goods, in the composition of their international trade, have also been responsible for distorting the relationship observed in developed countries (Pandit, 1986). Berry (1997) also endorses this view and attributes the phenomenon to rapid population increase in the early stages of demographic transition, which results in the tertiary sector becoming a major employment-providing area in cities due to slow growth of the secondary sector. Improvement in technology leads to shrinking job opportunities in the primary sector, necessitating migration to large cities, leading to 'false urbanisation'.

The sectoral shift also calls for a spatial shift as these activities require careful organisation and, thereby, an organising node or an urban centre; this leads to urban growth. It is not only accompanied by occupational shift, but also shows many other socioeconomic changes. Urban areas are, therefore, considered as foci of development and urbanisation crucial for accelerating the nation's economic development (Curie, 1966). At the global scale, there is a positive relationship between the level of urbanisation and development (Berry, 1961) and the phenomenon is replicated at the national level (Verma, 1989). It generates economic growth, but also creates inequalities of various kinds within populations and regions (Barkley et al., 1994; 1996). One, therefore, comes across a divide between these centres and the surrounding rural areas, which becomes sharper with advances in economy and city growth. Urban areas are often equated with high development and rural, with low.

The scenario has, however, undergone a change with the introduction of socialist ideas; the post-globalisation period, with its attendant requirements, is also making it increasingly necessary to reduce this divide. The rural-urban dichotomy is, thus, getting replaced by rural-urban transition (McGee, 1998). Although cities continue to act as attractive destinations for rural migrants, government investment in infrastructure facilitates rural-urban interaction. The whole phenomenon of rural-urban dynamics has thus undergone a change; rural communities have ceased to be rural without yet formally becoming part of an urban fabric (Brookfield et al., 1994). It is in this context that this paper attempts to understand the spatial pattern of association between urbanisation and other socioeconomic parameters related to development, as observed in the state of Maharashtra.

The Problem

The state of Maharashtra has shown a high level of urbanisation compared to India as a whole in all the Census years from 1961 to 1991, and its urban population has shown a steady increase over years in both absolute and relative terms (Table 1). The growth in the population of the Class I cities for the same period was 16.20 million in absolute terms. In other words, 83.59 per cent of the growth in the urban population was absorbed in Class I cities. One does not find any significant change in the spatial pattern of the relative levels of urbanisation in the state over years (Sita and Phadke, 1985; Phadke and Mukherji, 2004). The confirmation of this has been obtained by deriving a correlation matrix (Table 2). The correlation is not perfect and the value of the co-efficient has been going down with the passage of time, suggesting some slow changes over the years.

Table 1
Comparison of the Level of Urbanisation in Maharashtra with That in India, 1961-2001

Year	*Urban Population in Maharashtra (millions)*	*Urban Population (%) (Maharashtra)*	*Urban Population (%) (India)*
1961	11.163	28.2	17.8
1971	15.711	31.2	19.7
1981	21.994	35.0	23.2
1991	30.542	38.7	25.5
2001	41.020	42.4	27.8

Source: Census of India

Table 2
Correlation among Levels of Urbanisation in Different Census Years

	1961	*1971*	*1981*	*1991*
1961	1			
1971	0.988	1		
1981	0.973	0.990	1	
1991	0.934	0.957	0.976	1

In the light of the above, it was thought desirable to understand the forces that are still shaping urbanisation in the state of Maharashtra and study the extent to which development in other sectors is oriented to the phenomenon. When the analysis was carried out, relevant data for the year, 2001, had not yet been released and, hence, the work has been restricted to the year, 1991.

Conceptual Framework

The study attempts to understand urbanisation in relation to other socioeconomic dimensions of development. Although the list of indicators selected could have been long, there were constraints of availability of data. So, it was decided to select some key indicators. The level of urbanisation has been supplemented with contribution to the urban population of towns belonging to different size categories, such as large (Class I), medium (Classes II & III) and small (Classes IV, V & VI). The analysis with respect to size class of towns was expected to throw light on the nature of urbanisation, particularly in relation to the concentration or otherwise in any particular category. Selection of indicators 1-4 is thus justified (Table 3).

Table 3
Indicators Selected for Study

S. No.	*Indicator*
1	Level of Urbanisation
2	Contribution of Large Towns to Urban Population
3	Contribution of Medium Towns to Urban Population
4	Contribution of Small Towns to Urban Population
5	Percentage of Workers in Manufacturing Activities
6	Percentage of Workers in Construction
7	Percentage of Workers in Tertiary Activities
8	Percentage of Workers in Trade & Commerce
9	Percentage of Workers in Transport, Storage and Communication
10	Percentage of Workers in Other Services
11	Percentage of NSA under Irrigation
12	Percentage of Irrigated Area under Non-food Crops
13	Percentage of Villages with Bus Stop/Railway Station
14	Percentage of Villages with Telephone
15	Percentage of Villages with Higher Educational Institutions
16	No. of Dailies Published per Lakh Population
17	Percentage of Settlements with Regulated Markets.
18	Density of Population
19	Total Literacy
20	Female Literacy
21	Females among the Literate

Urbanisation is based on the changes in occupational structure from agriculture to non-agricultural activities; factory manufacturing, trade and commerce and other services are important occupations that promote the process. The construction boom is a natural concomitant. Indicators 5-10 have, therefore, been considered for analysis. Modernisation of agriculture has often led to development and irrigation water is the key input in the process. Proportion of net sown area under irrigation thus significantly contributes to development by permitting commercial agriculture in the form of raising non-food crops; this calls for processing and leads to structural changes in the economy, resulting in urbanisation and development. The indicators, 11-12, related to the agricultural sector have, therefore, been included.

Socioeconomic infrastructure is an essential requisite for structural changes, which lead to both development and urbanisation. Availability of transport, faster communication in the form of telephones and dissemination of information through mass media like daily newspaper, higher level education, including technical and professional, and marketing infrastructure are, therefore, essential. Indicators 13-17, which are related to these, thus deserve their inclusion.

The socioeconomic changes lead to demographic changes like rise in population density, literacy and growth of the population at a rapid rate. Growth rate was, however, not considered for the purpose of analysis as it stands over a duration of time, while all other indicators are related to only one point of time. Indicators (18-21) related to these have, therefore, been found essential.

Unit of Analysis

For a meso-level study region such as Maharashtra, district is too large a unit of analysis as it conceals micro-level variations, so it was decided to use taluka for the purpose. Taluka level urban population was analysed with respect to size class of towns along with other key indicators related to manufacturing, tertiary activities, commercial agriculture, socioeconomic infrastructure and demographic structure.

Data Base and Technique of Analysis

The study is based on secondary data on 21 selected indicators collected from the Census and district statistical abstracts for the state, for the year 1991 (Table 3). The technique of the principal components analysis has been employed, which made it possible to proceed on a mathematically sound basis to extract six factors from the initial statistical array. The factors reflect important linkages or regularities that could not be observed directly. The 'varimax' procedure of factor rotation made it possible to obtain orthogonal factors that do not correlate with one another. This has been done using the Minitab package; extraction was terminated when the variance extracted by a factor was less than 5 per cent.

Results of Factor Analysis

Six factors could extract 76.7 per cent of the variance and each of them has more than one variable associated with it (Table 4).

The first factor brings out the **urbanisation** dimension as it has a strong positive linkage with variables such as level of urbanisation, contribution of large towns to urban population and workers in activities such as other services, trade and commerce, transport and communication and manufacturing; this is indicated by their high primary loadings on to this factor. All these variables reflect on either the cause or effect of urbanisation and are thus city forming forces. Access to daily newspapers stands out prominently in its association with the factor, and workers in construction also contribute to the formation of this dimension by showing a significant association with it. One finds the considerable improvement in the association of workers in construction business with urbanisation over years (Phadke and Mukherji, 2003), and this is due to the new approach to development projects and the resulting construction boom to provide infrastructure, office space and

Table 4
Rotated Factor Loadings and Communalities

Variable	*F1*	*F2*	*F3*	*F4*	*F5*	*F6*	*Com. (h2)*
Urban Population	0.87	–0.19	–0.25	–0.20	0.03	–0.08	0.910
Large Towns	0.91	0.02	–0.08	0.05	–0.07	0.02	0.834
Medium Towns	–0.22	–0.07	0.03	–0.88	0.01	0.06	0.837
Small Towns	–0.20	–0.07	0.01	0.71	0.08	–0.18	0.588
Mfg. Workers	0.70	–0.19	–0.28	–0.07	–0.07	–0.24	0.671
Construction Workers	0.79	–0.34	–0.18	0.00	0.06	–0.04	0.774
NIA/NSA	–0.03	–0.05	0.02	–0.08	–0.82	0.15	0.700
GIA-NFC/GIA	–0.03	0.04	0.15	–0.11	0.48	0.60	0.624
Sett. with BS/Rly	–0.02	–0.21	–0.26	–0.01	–0.17	0.72	0.663
Sett. with Tel.	0.18	–0.17	–0.75	–0.12	–0.09	0.21	0.694
Sett. with H. Ed.	0.22	–0.04	–0.92	–0.03	0.05	–0.01	0.904
Dailies/Lakh Pop.	0.75	0.08	0.17	–0.04	–0.11	0.34	0.726
RM/100 Km2	0.19	–0.07	–0.17	–0.55	–0.07	–0.19	0.414
Pop. Density	0.32	–0.05	–0.90	–0.01	0.05	–0.05	0.924
Total Literacy	0.35	–0.86	–0.14	–0.08	–0.18	0.08	0.919
Females A. Lit.	0.13	–0.92	–0.02	0.06	–0.03	0.08	0.863
Female Literacy	0.35	–0.89	–0.14	–0.07	–0.14	0.03	0.961
HH Industry W.	0.10	–0.14	0.02	–0.07	–0.68	–0.10	0.502
Trade & Com. W.	0.81	–0.31	–0.37	–0.09	–0.02	0.23	0.903
Tr., St. & Com. W.	0.77	–0.35	–0.28	–0.05	0.02	–0.09	0.813
Other Services W.	0.82	–0.36	–0.26	–0.05	–0.06	0.00	0.874
Variance	5.78	3.02	2.88	1.71	1.49	1.23	16.099
Variance (%)	27.5	14.4	13.7	8.1	7.1	5.8	76.7

housing in large cities. Association of daily newspapers reflects on the rising demand for information and developments in related technology. Between 49 per cent and 82 per cent of the variation in the above variables is attributable to this factor, which accounts for 27.5 per cent of the total variance. Variables such as total and female literacy as also density of population have only their secondary loadings on to this factor. The contribution of such variables in the formation of the factor is not only small, it has also fallen over the years; this suggests that high density and literacy are no more a monopoly of urban areas (Goryacheva, 1977). Moreover, the indicator, 'Females among the Literate', shows very poor association and this is primarily because it is also affected by sex ratio (Phadke and Mukherji, 2001). The factor brings out the differentiation between the highly urbanised areas around metropolitan cities like Mumbai, Pune and other growing Class I cities at nodal locations, on the one hand, and the rest of the state, on the other (Figure 1). Some areas without large cities are also standing out because of their occupational character; for

Figure 1
Maharashtra: Factorial Ecology, 1991

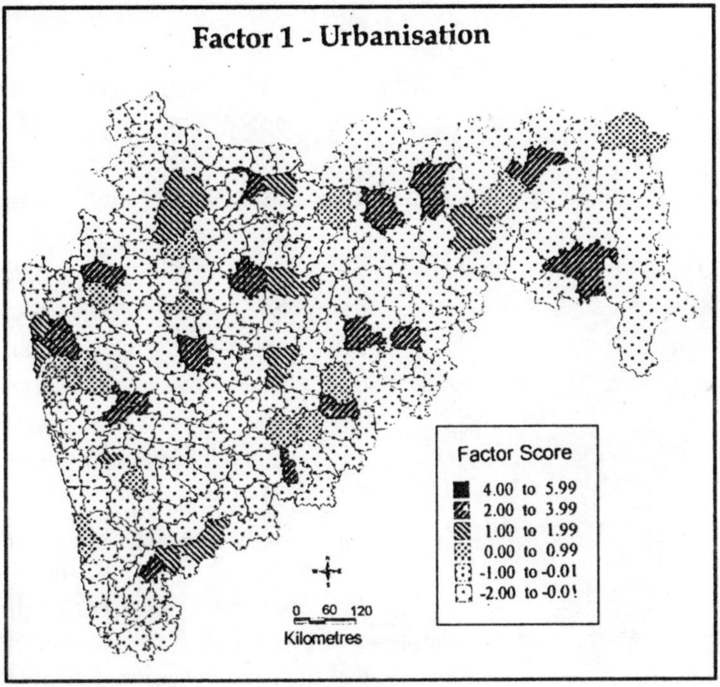

example, the taluka of Mahabaleshwar stands out because of its commercial importance, while Vasai has a higher factor score due to its location in the shadow of Brihanmumbai. Surprisingly, Brihanmumbai, which is expected to be at the highest level, considering its being the '*urbs prima*' of India, fails to do so primarily because the data is occupational in character, being collected at the place of residence (Sita and Phadke, 1982), and many people working in Mumbai have their place of residence outside its municipal limits. Among the other areas, which lie along the transport corridors, Marathwada stands out because of commercial activities (Department of Geography, 2003).

None of the indicators shows a significant positive loading on Factor 2, so it has to be technically labelled in a negative way. The negatively loading indicators are those related to

Figure 2
Maharashtra: Factorial Ecology, 1991

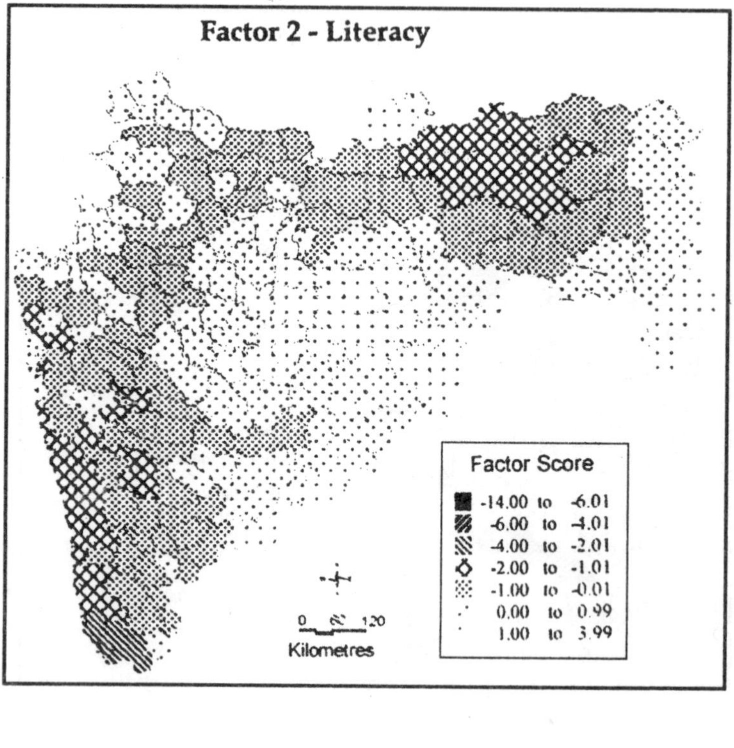

literacy; between 74 per cent to 84 per cent variation of the above variables is attributable to this factor, which extracts over 14 per cent of the total variance. The factor is thus labelled as the **literacy** dimension. Workers in the tertiary sector are also appreciably associated with the factor, indicating the need for literate people for conducting these activities. Workers in construction are also associated with the factor, partly because of construction projects such as the railways and, in part, due to the need for new housing and other infrastructure. The areas included here are mainly those belonging to Western Maharashtra and Central and Western Vidarbha (Figure 2). Differentiation between Konkan and *desh*, as also between North and South Konkan, are brought out prominently; this is also the case with the Central and other parts of Vidarbha. Areas of Marathwada, as also hill and drought prone areas, stand out by poor levels. Brihanmumbai makes a poor show because of its low score on females among the literate.

Factor 3 is also formed of negatively loading indicators, like proportion of settlements with telephone connection or higher education institutions and density of population. Between 56 per cent to 84 per cent variation of the above variables is attributable to this factor, which extracts nearly 14 per cent of the total variance. Association between density and infrastructure is primarily because of the fact that better infrastructure is available only in the areas of population concentration, where it would be usually in demand. Surprisingly, the indicator related to access to transport is not intimately related to this factor, which reflects on the availability of infrastructure *sans* transport and the population served by it. It is labelled the **infrastructure sans transport** dimension. Workers in trade and commerce are also appreciably linked to this factor, as the activity, being demand specific, is expected to flourish in densely populated areas. Spatially, areas of *desh* and some pockets in Marathwada stand out (Figure 3). While Mumbai and Pune are at the highest level, this is not the case with Nagpur; the latter stands out at the

Figure 3
Maharashtra: Factorial Ecology, 1991

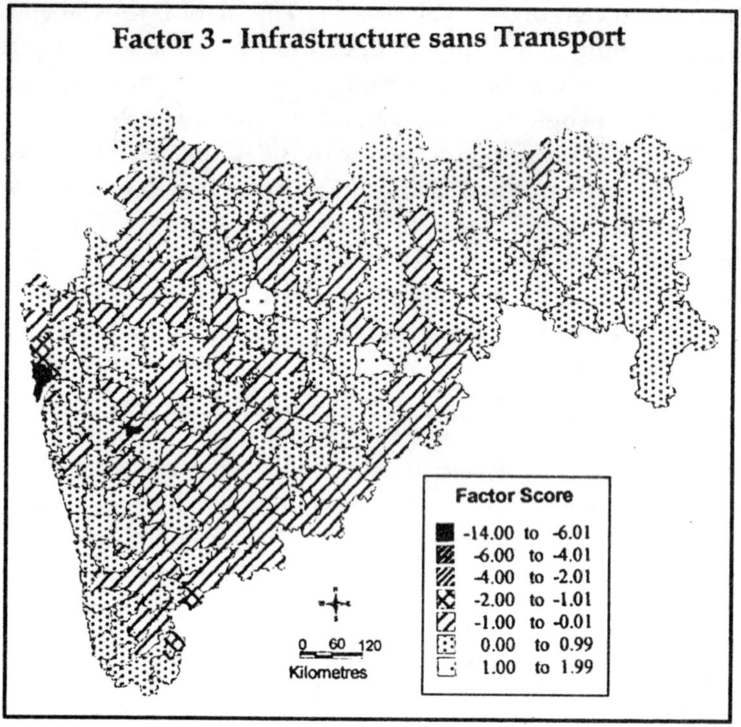

same level as the other areas. The differentiation between Konkan and *desh*, on the one hand, and Western and Eastern Vidarbha, on the other, is obvious. Some of the remote hilly areas of the Western Ghats show better availability of infrastructure because services like telephones were introduced there as an administrative exigency.

Medium-small town polarity is the fourth dimension; access to regulated markets goes with the medium towns. In fact, the availability of infrastructure in medium towns promotes them as locations for regulated markets. The level of urbanisation is only poorly related to this factor, suggesting that these urban centres primarily fill the void created by the

Figure 4
Maharashtra: Factorial Ecology, 1991

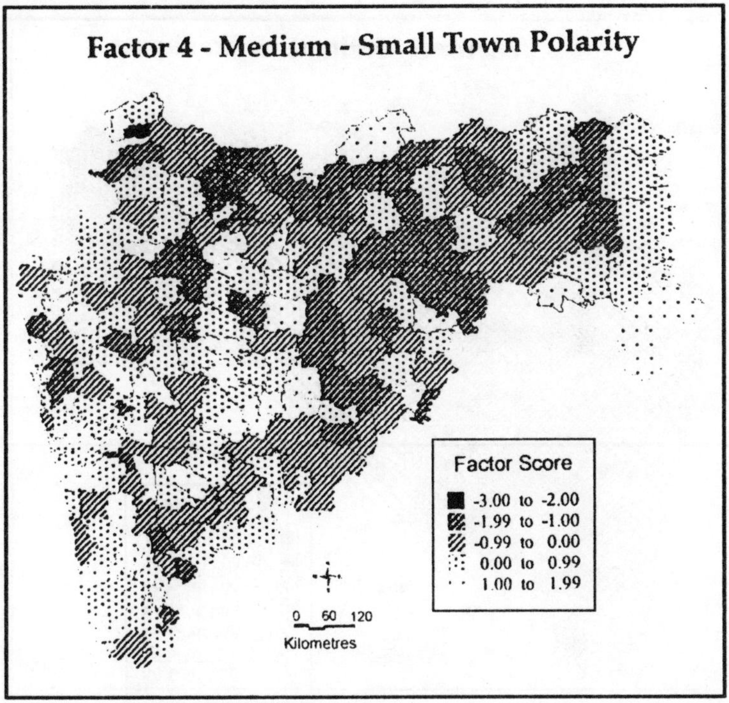

absence of large towns, which primarily contribute to the level of urbanisation. Areas of Vidarbha and Marathwada, alongside transport corridors, but outside the major city talukas of Akola, Amaravati and Aurangabad, stand out as areas of medium towns (Figure 4); this is also true of Western Maharashtra, but Pune city taluka is an exception to this, as it has two medium towns incorporated in it besides Pune City, as also regulated markets. The hilly and marginal areas are characterised by small towns.

Irrigation and household industry constitute the fifth dimension, which is labelled the **irrigation** factor. Irrigation provides agricultural raw materials by way of cash crops like

Figure 5
Maharashtra: Factorial Ecology, 1991

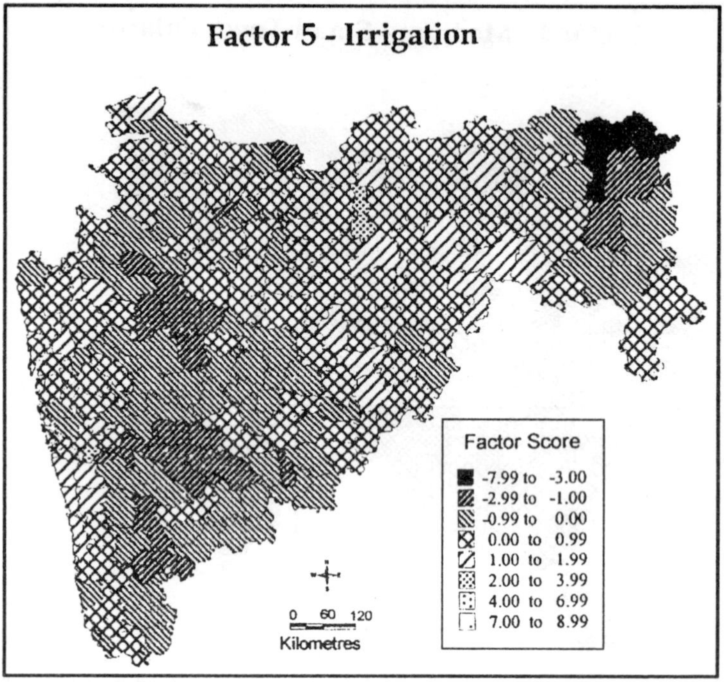

oilseeds and tobacco for processing in household workshops. This is brought out by fairly high secondary loading of irrigated non-food crops on to this dimension. The irrigated areas of Western Maharashtra and Eastern Vidarbha stand out prominently, the latter primarily on account of its position in household industries. Marathwada and Konkan are also being slowly brought into the fold (Figure 5).

Transport infrastructure forms the sixth dimension and areas under irrigated non-food crops shows a strong loading on to it; transport is an essential requirement for quick movement of the produce of such crops. The spatial pattern brings out the contrast between Vidarbha and rest of the state; areas of Eastern

Figure 6
Maharashtra: Factorial Ecology, 1991

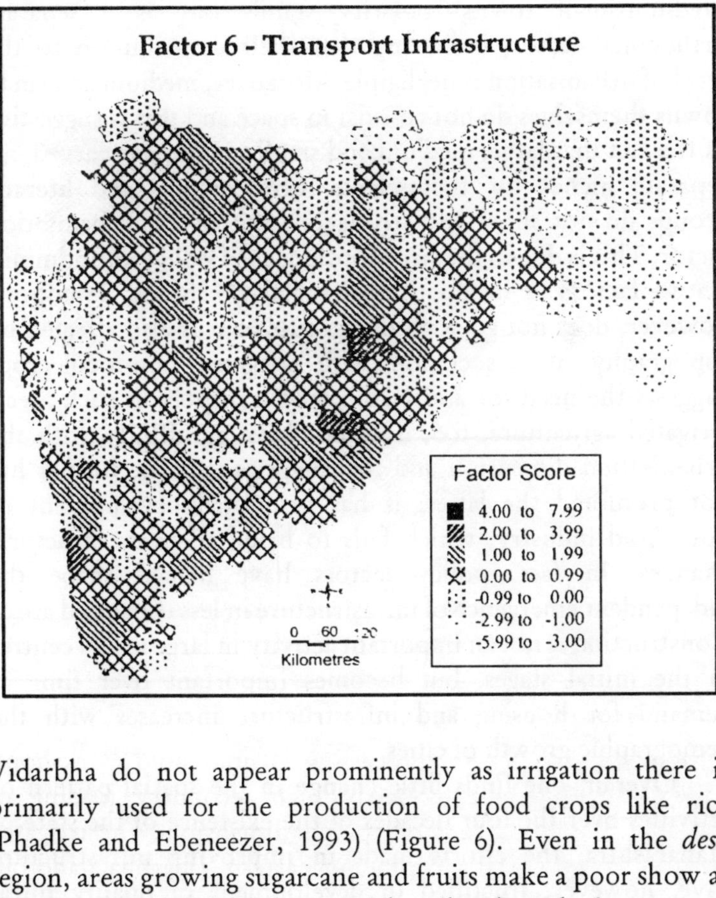

Vidarbha do not appear prominently as irrigation there is primarily used for the production of food crops like rice (Phadke and Ebeneezer, 1993) (Figure 6). Even in the *desh* region, areas growing sugarcane and fruits make a poor show as they are considered food crops. The oilseed producing areas of Marathwada, the Tapi-Purna valleys and South Konkan have a better representation.

Concluding Remarks

The analysis so far clearly reveals that urbanisation is related only to workers in some important non-agricultural categories, like factory manufacturing, construction and different tertiary activities. The level of urbanisation is strongly related only to the contribution to urban population by large cities. Medium-small towns' polarity stands out as a separate orthogonal factor, indicating that their contribution to the level of urbanisation is negligible. Moreover, medium and small towns themselves do not co-exist in space and this is suggestive of the fact that large, medium and small towns have carved out separate niches for themselves. Infrastructure and literacy groups do not show any strong relation to the urbanisation factor. This is due to the strong influence of policy and administrative factors in localisation of social overheads. Transport, however, does not go with this group, as it is also affected by topography; its association with irrigated non-food crops suggests the need for and viability of transport in such areas. Irrigated agriculture, too, does not have an affinity with the urbanisation dimension and this suggests that the former has not promoted the latter; it has promoted development of household industry, which fails to bring adequate structural changes. In fact, policy factors have helped shape the independent emergence of infrastructure in less urbanised areas. Construction is not an important activity in large urban centres in the initial stages, but becomes important over time as demand for housing and infrastructure increases with the demographic growth of cities.

Overall, one finds little change in the spatial pattern of activities over the four decades of the existence of the state of Maharashtra; the efforts made in improving infrastructure have, however, fructified in development of quality infrastructure in less developed areas. Investment in infrastructure has facilitated rural-urban interaction and thereby changed rural-urban dynamics. The advent of the telematics revolution and shrinkage of cost and time distances have brought in radical change in rural-urban relations. Today, one does not observe a sharp rural-urban dichotomy as in the past, but transition still exists. These developments have not changed the relative pattern of prosperity substantially and there continues to be differentiation in income levels. Cities still pull rural dwellers, although the push factors are becoming weaker day by day. However, the fact remains that rural communities have ceased to be rural in the traditional sense, although for technical purposes, they have not still become a part of an urban fabric. This augurs well and it is necessary to take advantage of this opportunity and the momentum generated thereby in these semi-urban regions. But the uplift of these regions requires development of a proper perspective, which would need a rural orientation to make the community feel that it is an equally respected partner in the development process.

Acknowledgement

The findings of this paper are based on the project 'Urbanisation and Development Scenario of Maharashtra', funded by the I.C.S.S.R., New Delhi, during 2001-04. The authors thank the organisation for the same. The authors also acknowledge the help received from the Department of Statistics, University of Mumbai, in making factor analysis possible.

References

Barkley, D.L., and Henry, M.S. (1996): 'Big City Lights and Small Towns: Conflicts and Confluence', Centre for Community Economic Development, CEN Newsletter No. 237.

Barkley, D.L., Henry, M.S., and Bao, S. (1994): "Metropolitan Growth – Boon or Bane to Nearby Rural Areas?', *Choices*, Fourth Quarter, 14-18.

Bell, D. (1973): *The Coming of Post-industrial Society: A venture in Social Forecasting*, Basic Books, New York.

Berry, B.J.L. (1961): 'Basic Patterns of Economic Development', in N. Ginnsberg (ed.), *Atlas of Economic Development*, The University of Chicago Press.

Berry, B.J.L. (1997): 'Reversal of a Relationship: A Note on the Tertiary Sector', *Urban Geography*, 18(5), 460-7.

Brookfield, H., et al. (1994): 'The Challenge of Southeast Asia's Rapid Motorisation: Kuala Lumpur, Jakarta, Surbaya and Manila in an International Perspective', Paper presented at the Asian Studies Association of Australia Biennial Conference, Perth.

Clark, C. (1940): *The Conditions of Economic Progress*, Macmillan, London.

Curie, L. (1966): *Accelerating Development*, McGraw Hill, New York.

Department of Geography (2003): 'Marginal Towns of Maharashtra, 1961', M.A. Part II Project Report, Department of Geography, University of Mumbai.

Fuchs, V.R. (1968): 'The Service Economy', National Bureau of Economic Research, New York.

Gong, H. (2002): 'Growth of Tertiary Sector in China's Large Cities', *Asian Geographer*, 21(1-2), 85-100.

Goryacheva, A.M. (1977): 'Factor Analysis of the Growth of India's Largest Cities and Differentiation of Economic Functions', *Soviet Geography, Review and Translation*, XXVIII(6), 384-95.

Kellerman, A. (1985): 'The Evolution of Service Economy: A Geographical Perspective', *Professional Geographer*, 37(2), 133-43.

McGee, T.G. (1998): 'Globalisation and Rural-Urban Relations in the Developing World' in Lo Fu-chen and Yue-man Yeung (eds), *Globalisation and the World of Large Cities*, United Nations University Press, 471-96.

Pandit, K. (1986): 'Sectoral Allocation of Labour Force with Development and the Effect of Trade activity', *Economic Geography*, 62, 144-54.

Pandit, K., and Casetti, E. (1989): 'The Shifting Pattern of Sectoral Labour Allocation During Development: Developed versus Developing Countries', AAAG, 79(3), 329-44.

Phadke, V.S., and Ebeneezer, S. (1993): 'Is Irrigated Agriculture in Maharashtra Secured against Drought?', Occasional Paper, WRC, I.C.S.S.R., Mumbai.

Phadke, V.S., and Mukherji, D. (2001): 'Spatial Pattern of Population Distribution and Growth in Brihanmumbai, 2001', *Urban India*, XXI(2), 111-33.

Phadke, V.S., and Mukherji, D. (2003): 'Urbanisation and Development in Maharashtra, 1961-91: Some Observations', *Transactions*, IIG, 25 (1&2), 116-24.

Phadke, V.S., and Mukherji, D. (2004): 'Urbanisation and Development Scenario of Maharashtra, 1961-91', Report of the I.C.S.S.R. Funded Project, Department of Geography, University of Mumbai.

Sita, K., and Phadke, V.S. (1982): 'Use of Occupational and Employment Data in Intra-Urban Analysis – A Comparative Assessment', *Geographical Mozaic* (Prof. K.N. Bagchi Felicitation Volume), Kolkata, 667-73.

Sita, K., and Phadke, V.S. (1985): 'Urbanisation in Maharashtra State, 1971-81' in N. Prasad et al., *Modern Geographical Concepts* (Prof. M.R. Chaudhari Felicitation Volume), Kolkata, 87-96.

Verma, S.S. (1989): *Urbanisation and Regional Development in India*, Chugh Publications, Allahabad.

8

Decolonisation of the Administrative Organisation of Space in Punjab since Independence

Gopal Krishan and Surya Kant

Introduction

The world is characterised by a variety of spatial organisations: political, economic, religious, juridical and administrative, to list some. Administrative organisation of space refers to the overall arrangement and internal divisions of political areas into a hierarchy of administrative units to manage public affairs. These units form the basis of designing and executing public policy, have a strong bearing on the pattern of supply and utilisation of public services, and are influential in directing and standardising the life of the people (Garner, 1975; Beaujeu-Garnier, 1976; Bennett and Chorley, 1978).

It is estimated that there exist in the world over two hundred national level administrative territories, about three thousand regions of position analogous to the French departments or the English counties, about forty to fifty thousand units comparable to districts, and about three hundred thousand basal units in the nature of local communities (Blazek, 1976). All of them offer a great variation in their evolutionary process, organisational set-up, spatial attributes, obligations and functions.

On the face of it, any administrative organisation of space seems to be a mechanism for a convenient management of large, heterogeneous and disparate units. In spirit, this organisation is deeply rooted in the whole philosophy of government and society in any region. It reflects the basic values, such as efficiency, justice, democracy, community and development, of the people in these areas (Krishan, 1990). In real life situations, however, the considerations of convenience of management and efficiency outweigh all other factors.

Administrative Areas

The most critical factor which distinguishes administrative areas from each other is the degree of functional autonomy/ sovereignty they enjoy. These can, accordingly, be classified into four types (Humes and Martin, 1961): (i) supra-sovereign, for example, the United Nations and the European Union; (ii) fully sovereign, such as the United States of America and India; (iii) quasi-sovereign or the federated state, such as California in the United States of America and Punjab in India; and (iv) infra-sovereign or the local authority area, for instance, Orange County in California and Ludhiana district in Punjab. These four types obviously differ on the spatial scale of their operations.

Another way to classify administrative areas is to categorise them on the basis of their 'function': general purpose administrative areas, special purpose authorities and regional authorities. The general purpose administrative areas are in the nature of all-purpose units, such as districts in India. These provide a multitude of administrative services. The special purpose administrative areas are meant for specific functions, such as school districts in the United States of America, health units in Botswana, and milk marketing areas in Canada

(Massam, 1972). The formation of special purpose authorities is essentially a sectoral division of the general purpose administrative areas. Regional authorities are constituted to manage and provide some higher order service, specific to a particular function or area. Electricity boards, transport corporations and river valley authorities are some such arrangements. These tend to be a hybrid between the general purpose and special purpose administrative areas. Generally, these require coordination among a number of general purpose administrative areas (East and Prescott, 1975).

A periodic reform of all administrative maps is deemed imperative (Krishan, 1988). This is the practice in practically all countries of the world. In developed countries such as the United Kingdom, Germany and Sweden, the lowest level administrative units are being consolidated into bigger ones. The latter are rated as economically more efficient. In developing countries such as India, the process is in reverse. The existing administrative units are divided into smaller ones. The latter are deemed as more manageable.

In the United Kingdom, a Royal Commission on Local Government is periodically appointed to advise on the necessary revision of the county and district boundaries. Questions like the optimal area/population size of these administrative units, the desired number of tiers amongst them and the manner of implementation of the recommended changes are examined and debated. All approved reforms of the administrative maps come into effect on a single date scheduled for the purpose.

Such an administrative culture is yet to mature in India. Administrative area changes are carried out in an ad hoc and piecemeal manner. The concerned reports are treated as strictly confidential, open neither to the public eye nor to public debate. Any academic exercise on the matter is grossly constrained by non-availability of authentic material.

The Research Question

Taking a cue from these propositions, this paper proposes to reflect on the evolving scene of administrative organisation of space in Punjab during the post-Independence era. The basic question is: To what extent does the changing administrative map of Punjab since Independence represent the process of decolonisation in terms of the theory and practice of administration? The reference frame is set by the basic tenets of British administration in India. These included precise territoriality, centrality, legality, rationality and impersonal bureaucracy. The contours of the evolved scene will be referred to these parameters. A comparative view will be taken of the colonial and post-colonial periods to draw appropriate inferences.

The paper is organised into two sections. The first describes the administrative scene of Punjab during the British days, and the second deals with the nature of general purpose administrative areas, special purpose authorities and regional/sub-regional authorities during the post-Independence era. An attempt will be made to identify phases in the recasting of the territorial arrangements for administration.

Colonial Administration Landscape

Punjab became a part of British India in 1849. Soon after, in 1875, an uprising broke out against the British. It was suppressed and followed by radical changes in administrative policy.

Through an act of British parliament, the authority to rule India was transferred from the East India Company to the British Crown. A clear distinction was made between the loyalists and rebels during the uprising. In a major administrative reorganisation bid, the former Mughal Province of Delhi, then forming a part of Northwest Provinces, was transferred to Punjab. A number of princely states, such as Patiala, Jind and Kapurthala, were not only allowed to maintain their former status but also awarded land grants. Punjab had two segments, the British ruled territory and the princely states.

Map 1
Pre-Independence Punjab: Partitioned in 1947

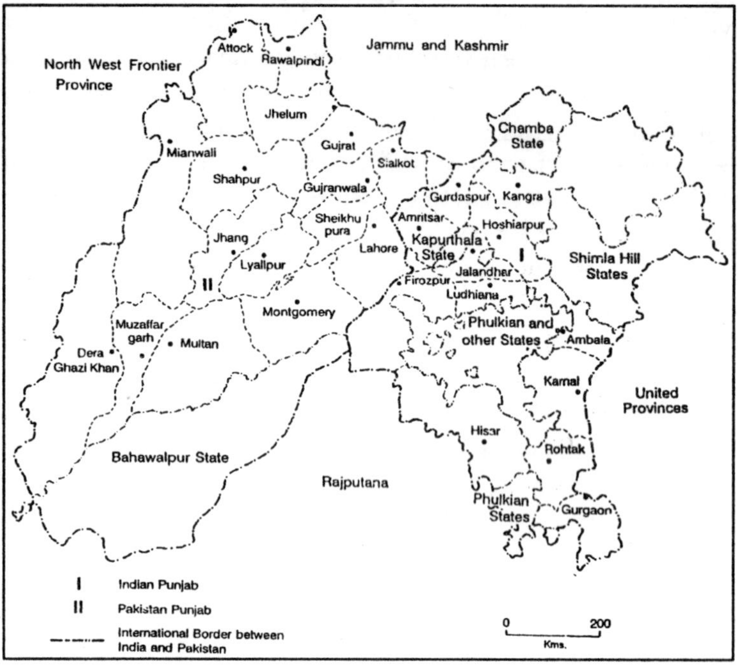

In 1901, the Northwestern part of Punjab was constituted as a separate North West Frontier Province. In 1912, Delhi was detached to form a new province under a chief commissioner. Thereafter, no significant territorial change took place in the boundaries of Punjab till partition in 1947 (Map 1).

The Punjab of 1901 was divided into 27 districts, grouped into five divisions. In 1904, two new districts of Attock and Lyallpur were carved out. Besides, there were 43 native states of varying size and importance scattered over different parts of Punjab. The British Punjab accounted for 70 per cent of the total area and the native states, 30 per cent (Table 1).

An average district of Punjab spread over an area of 8,632 km^2. Districts were divided into tehsils/sub-collectorates. There were 104 tehsils/sub-tehsils in Punjab.

Table 1
Punjab: Area, Population, Settlements and Administrative Units, 1901-1996

Year	*Area (in km²)*	*Population (in million)*	*Number of*	
			Rural & Urban Settlements	*Administrative Units*
1901	389,036	26.88	47,256	32 Districts 43 native States
1911	353,095	24.19	44,574	29 Districts 15 Punjab States
1921	354,688	24.95	45,409	29 districts 15 Punjab states 28 Simla hill states
1931	354,400	29.13	51,224	29 Districts 15 Punjab States 28 Simla Hill States
1941	359,179	35.23	52,644	29 Districts 16 Punjab States 27 Simla Hill States
1951	96,809	12.64	15,277	13 Districts
1961	122,005	20.35	22,879	19 Districts
1971	50,362	13.55	12,450	11 Districts
1981	50362	16.79	12,450	12 Districts
1991	50362	20.19	12,915	12 Districts
1996	50362	22.29*	12,915	17 Districts

Source: Census of India, 1901 to 1991
*Estimated

A tehsil was primarily a revenue administrative unit. It was subdivided into *kanungo* circles. These were comparable to the Mughal *parganas/mahals*. Each *kanungo* circle had a police *thana*. The latter, comprising a group of villages, was the lowest unit of police administration. On the other hand, a *patwar* circle functioned as an intermediate tier between a *kanungo* circle and the village for revenue administration.

As such, the British borrowed a number of administrative practices, including terminology, from the Mughals. There were some notable departures as well: (i) new tiers of divisions and sub-divisions were introduced for stronger administrative control; (ii) revenue administration was brought down to more local levels of spatial hierarchy; and (iii) the police administration was also spatially deepened, not only to have a firm grip on law and order, but also to make revenue collection more effective (Kant, 1988).

Contemporary Punjab

At the dawn of Independence, when the country was rejoicing over its newly acquired freedom, Punjab was reeling under the worst communal riots in its history. In 1947, the province was partitioned between India and Pakistan along religious lines. Of its 3,59,179 km^2, five divisions and 29 districts, only 152,649 km^2, two divisions and 13 districts were left with Punjab (East), renamed Punjab (India) on 26th January 1950.

The Punjab native states presented a somewhat different picture. Eight of them, namely Patiala, Kapurthala, Jind, Nabha, Faridkot, Malerkotla, Nalagarh and Kalsia, were constituted into a union, called PEPSU, in 1948. These covered an area of 26,156 km^2 and were arranged into eight districts and 25 tehsils.

In another move, the former Punjab hill states were constituted into Himachal Pradesh in the same year. The new unit was a centrally administered territory. It was divided into four districts and 23 tehsils. The hill state of Bilaspur, which became the scene of Gobind Sagar associated with the Bhakra Nangal Project, was made a part of Himachal Pradesh at a later stage, in 1954.

Subsequently, Punjab underwent two major reorganisations. In 1956, PEPSU was merged with Punjab on the recommendation of the States Reorganisation Commission. Later on, in 1966, under the Punjab Reorganisation Act, a Punjabi speaking state was carved out of the erstwhile Punjab.

The second reorganisation reduced the state to two-fifths the size it had been after the first reorganisation in 1956. Now it covers an area of only 19,445 sq. miles (50,362 km^2). This is just one-seventh of its size before Independence. Though much reduced, Punjab was now linguistically homogeneous and structurally compact. It was divided into 11 districts, including Ropar, created as a byproduct of the reorganisation.

It follows that in the initial phase following Independence, the main thrust was on recasting the administrative organisation of space inherited from the British. A notable innovation can not be missed. This was the formation of the development block as the basic unit for rural development. It represented a new philosophy in area administration. The message went home that the administration was now trying to shift its focus from the maintenance of law and order and collection of land revenue to matters of development. The process of formation of development blocks in Punjab started in 1952. By 1963, the entire state had been covered by 117 blocks.

General Purpose Administrative Areas

At present, Punjab is organised into four divisions, 17 districts, 70 subdivision/tehsils, 55 sub-tehsils and 137 development blocks (Figure 1; Table 2; Map 2). At the time of its formation in 1966, the State had two divisions, 11 districts, 37 sub-divisions/tehsils, 13 sub-tehsils and 117 development blocks. A tendency towards creation of new administrative units at all levels is unmistakable. The process gained momentum in the mid-1980s. A spurt in the formation of new districts, sub-divisions and development blocks was experienced in the mid-1980s. The number of districts went up 1.5 times, that of divisions and sub-divisions/tehsils two times, and that of sub-tehsils four times during the last 30 years. By comparison, the number of divisions, as also of districts in the British Punjab, remained constant for the period, 1912-47.

Figure 1
General Purpose Administrative Areas in India (A Six-tier Hierarchy)

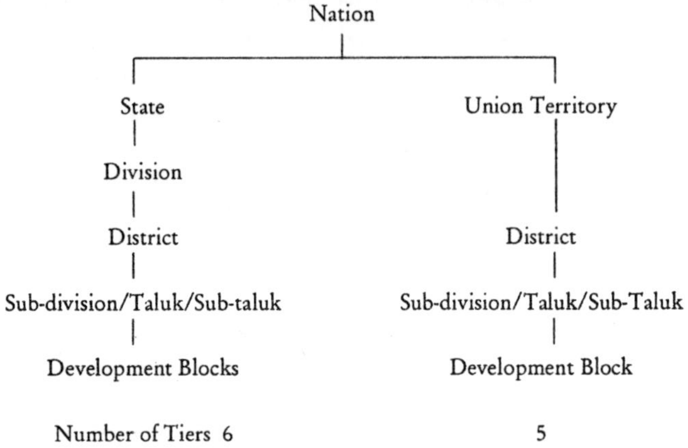

Table 2
Punjab: General Purpose Administrative Areas, 1966-96

Type	*Number in the Year*		
	1966	*1985*	*1996*
Divisions	2	3	4
Districts	11	12	17
Sub-divisions/Tehsils	37	46	70
Development Blocks	117	118	137

The Division

In the chain of general administration, the division represents the first order organisation of the state. Its origin in India goes back to the British days when, in 1829, for the first time, an intermediate authority between the district collector and the provincial government was created in the then Bengal province. In essence, the division emerged through an amalgamation

Map 2

Punjab: Administrative Divisions (General Purpose Administrative Areas) 1996

process. The practice was subsequently followed in other provinces. Every province in India, barring Madras, was organised into divisions.

The division was initially created as a revenue unit. The divisional commissioner not only heard appeals from subordinate revenue authorities but also acted as a supervisory and coordinating authority over collectors and their subordinate revenue officials. Gradually, additional administrative functions and powers were delegated to him. A division got eventually transformed into a general purpose unit.

The advisability of creating an intervening authority between a province and districts was questioned even during the British period. The debate intensified after Independence. It was argued that the British adopted this system because it helped them to centralise their authority. It could now be changed to a more democratic pattern (Sharma, 1971). This was seen as leading to a direct link between a district and the state, obviating the necessity of having any intermediate irritant. Madhya Pradesh, Maharashtra, Rajasthan and Gujarat abolished the divisions in 1948, 1950, 1961 and 1964, respectively. The territories occupied by Madhya Pradesh, Maharashtra and Rajasthan, however, reverted to the old system after some time. Punjab retained the traditional set-up all through. It was divided into four divisions. One is not sure about the relevance of this practice for a small state like Punjab.

The District

The district has been the key unit of local administration in India since the historic past. The Mauryas organised the district as the focal point of local administration more than two millennia ago. Akbar followed the same practice during mediaeval times. The British also favoured an administrative system that centralised powers and responsibilities in one hand at the district level. At the head of every district, it appointed an officer who, in the eyes of most of its inhabitants, was 'the government' (Shukla, 1976).

Independent India, too, continued with the tradition of having the district as the pivot of the local apparatus of public administration. The district administration in India is non-elective. An officer from the Indian Administrative Service is placed as its chief. The person is known as deputy commissioner in Punjab, district magistrate in Uttar Pradesh and collector in Rajasthan. This reflects the difference in the perceived role of a district head as custodian of law and order or as dispenser of justice or as collector of revenue.

There are 17 districts in Punjab (Map 3). This gives an area of about 3,000 km^2 per district. The comparable figure for India is around 6,000 km^2. An average district in Punjab is just half the area of its counterpart in the rest of India.

The Sub-division

The Mughals had a tradition of dividing a district (*sarkar*) into a number of *parganas* or *mahals* (sub-divisions) in India. This was primarily for revenue purposes. The British adopted the same practice. They, however, soon substituted the *parganas* with talukas or tehsils. Some parts of a district, depending upon their size or revenue or strategic location, were given the status of sub-division while the remaining part was divided into tehsils. Some tehsils were further divided into sub-tehsils. The system continued after Independence and under it sub-divisions enjoyed higher status than tehsils. Now almost all district level departments have their field offices at the sub-divisional headquarters.

Currently, Punjab is divided into 70 sub-divisions. This number has grown from 37 in 1966. Of the 33 new additions, nine came during 1966-85 and the remaining 24 during 1985-96. An average sub-division is now less than 800 km^2 in area. This is a size smaller than that earmarked for a block (1,300 km^2) initially.

Maps 3 & 4

Punjab: Spatial Organisations for Administration

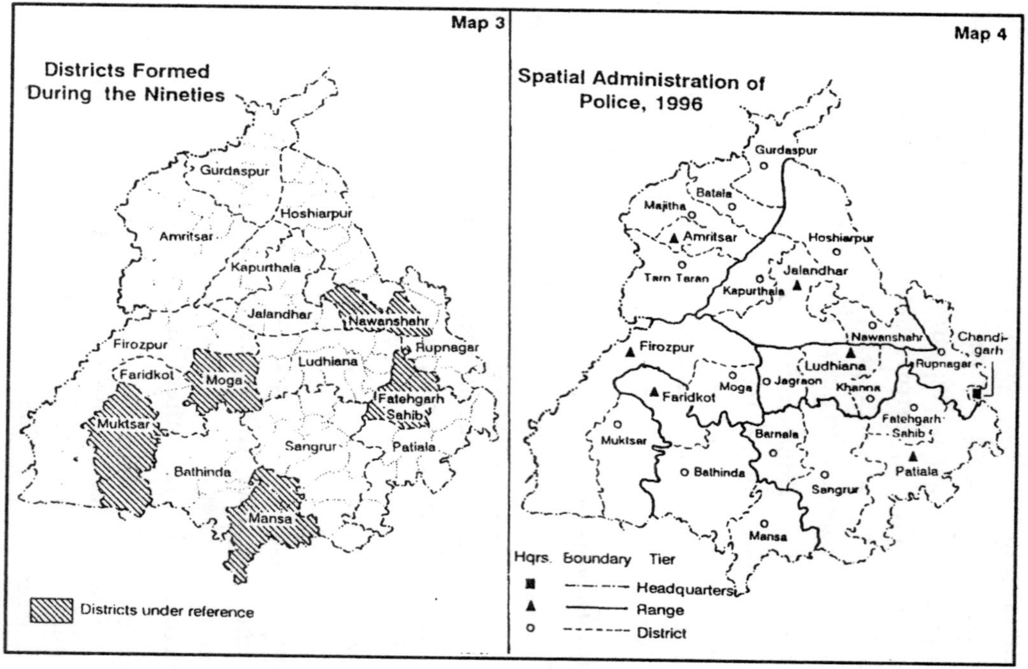

The Development Block

The block as an administrative tier in the district administration was an outcome of the Community Development Programme and the National Extension Service Scheme initiated on 2nd October 1952 and 1953, respectively. These were intended as peoples' programmes with government participation and not as government programmes with peoples' participation for rural uplift. Urban areas were kept out of the purview of development blocks.

As noted above, Punjab had 117 development blocks in 1966. Now the number is 137. Of the 20 new development blocks created during 1966-96, 19 came into existence after 1985.

Special Purpose Authorities

For the purpose of our understanding, these can be first classified on the basis of the spatial scale of their operations. In the case of India, these can be classified into three types: (i) those organised only at the national level, such as, railways, telecommunications and defence (Figure 2); (ii) those organised at both the national and state levels, including education, health and civil supplies, known as concurrent subjects (Figure 3); and (iii) those organised only at the state level, including the police, revenue and rural development, known as state subjects (Figure 4).

Special purpose authorities can be further grouped into six categories on the basis of their specific functions: (i) provisions of personal services, such as education, health and communications; (ii) management of environmental services, such as transport, housing and planning; (iii) sectoral administration as that of agriculture, industry and forestry; (iv) conduct of elections; (v) defence, and (vi) data collection, as on population and agriculture.

Figure 2
Spatial Organisation of Special Purpose Authorities Organised at the National Level

Spatial Coverage	Postal Services	Justice	Conduct of Elections	Census Operations
National and Inter-state	Board	Supreme Court	Election Commission Office	Registrar General Office
State and Intra-state	Circle	-	-	-
	Region	High Court	State Election Office	State Census Directorate
	Division	-	-	Census Division
	District/Head Post Office	District Court	Election District	Census District
	Post Office	Sub-divisional Court	-	Census Charge
	Sub-post Office		-	Census Circle
		-	-	Census Block
Number of Tiers	7	4	3	7

The national government undertakes a number of activities for the country as a whole. This is accomplished through the formation of a number of special purpose authorities. Among these, post and telecommunications, defence, administration of justice, institutional finance, conduct of elections and collection of a variety of demographic, climatic and economic data are the important ones.

In addition to the 96 specific subjects on its list, the national government shares responsibilities with the state governments in another 47 subjects. Included here are education, health, environment, planning, social welfare and scientific research.

The spatial organisation of special purpose authorities of the Central and state governments combined normally

Figure 3

Spatial Organisation of Administration of Special Purpose Authorities Jointly at the National and State Levels

Spatial Coverage		Education				Medical		
		Ministry of Education				Ministry of Health and Family Welfare		
National and Inter-state Levels								
	University Grants Commission	Region		Indian Council of Medical Research		Region		
		Ministry of Education				Ministry of Health and Family Welfare		
	University							
		Directorate	Directorate	Directorate	Directorate	Directorate	Directorate	Directorate
State and Inter-state Levels		College	—	—	—	College	—	College
		—	Circle	—	—	—	—	College
		—	District	District	District	—	Civil Hospital	District
		—	—	—	—	—	Hospital	—
		—	—	Block	Sports Club	—	Primary Health Centre	—
		—	—	—	—	—	Dispensary	Rural Dispensary
No. of Tiers		4	3	3	2	3	7	3

Figure 4
Spatial Organisation of Administration of Special Purpose Authorities at the State Level

Sector	Police	Agriculture	Industry	Forestry	Land Revenue	Rural Development	Soil Conservation
	Headquarters	Directorate	Directorate	Head Office	Head Office	Directorate	Head Office
	Range	-	-	Circle	Division	Division	Circle
	District	District	District Center	Division	District	District	Division
	Circle	-	-	Range	Tehsil	-	Sub-division
	Police Station	Block	Block Office/ Industrial Estate	Sub-range	Kanungo Circle	Block	-
	-	-	-	Beat	Village	-	-
Number of Tiers	6	4	3	7	7	5	4

corresponds with that of the general purpose administrative areas. The Central government looks after the higher level of a particular service, while the activity sphere of the state government is much more comprehensive. The latter runs the same service through a larger number of tiers of administrative areas.

There are sixty-six items listed as state subjects in the Constitution of India. Important among these are maintenance of law and order, agriculture, industry, forestry, revenue and cooperation. The state subjects are not fully protected from takeover by the Central government under special circumstances.

Law and order is a state responsibility and it is assured through police administration. In practically all states, including Punjab, the police administration has a six-tier spatial hierarchy: police headquarters, range, police district, circle, *thana* and post. The state is organised into six ranges and 22 police districts (Map 4). By comparison, the number of divisions and districts for general administration is four and 17, respectively. The disturbed law and order situation during the 1980s impelled a more intense spatial arrangement for the police administration.

Agriculture, industry and forestry are among the most important sectors at the state level. Their administration involves policy making and programme implementation. The state government often associates itself with the Central government in executing schemes relating to these sectors.

The administration of agriculture in Punjab is four tiered: headquarters, district, block and focal point (Map 5). In 1972, the agriculture administration was reorganised to cover new responsibilities and remove existing anomalies. Major rationalisations included (i) enhancement in the powers of district level officers, (ii) decentralisation of agricultural administration to the block level; and (iii) greater coordination of the agricultural administration with that of other special purpose authorities, such as rural development, Panchayati Raj and cooperation. In addition, agricultural problem regions, such as the border belt, the foothill (*kandi*) zone, and the *bets* (flood plains) were identified and given special attention, signifying area focus.

Punjab cannot take an equal pride in industrial development as in agriculture. Its border location with Pakistan and an almost complete absence of mineral resources explain this state of affairs. The state has, however, made commendable progress in agro-based and agro-oriented industries.

The administration of industry in Punjab is again three tiered: headquarters, district centre, and block office/industrial estate. Block offices serve rural areas and industrial estates are meant for towns. 'Industrially backward' and 'no industry' districts have been identified at the Central level to receive special assistance (Map 6).

Although forests cover only 1.5 per cent of the state's total area, yet their spatial administration is organised in seven tiers: headquarters, circle, division, range, sub-range, block/round and beat. Special efforts are being made to promote forestry in the foothill districts of Gurdaspur, Hoshiarpur and Rupnagar under the Kandi Watershed and Area Development Project assisted by the World Bank. Great importance is being attached to social forestry programmes that encourage plantation of trees in village commons. This has made forestry a spatially widespread activity for which additional powers have been delegated to the lower levels of the administrative hierarchy.

Historically, land revenue has been the major source of state income in India. The civil administration at the local level was virtually synonymous with revenue administration. In the post-Independence era, however, its importance to state income has declined greatly. Land revenue administration, however, continues to be one of the most elaborately organised systems in seven tiers: headquarters, division, district, tehsil, *kanungo* circle, *patwar* circle and revenue *mauza* (village). It works in close cooperation with the general purpose administration and their spatial organisation largely corresponds to each other. Development blocks are, of course, outside the purview of

Maps 5 & 6

Punjab: Spatial Organisations for Administration

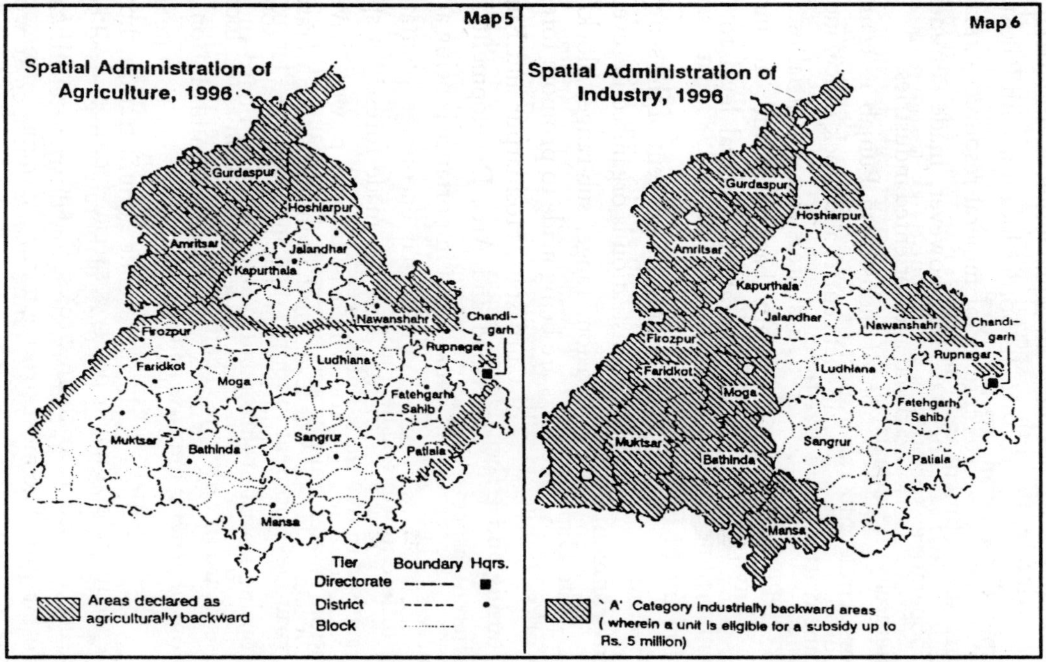

revenue administration and *kanungo* circles are not general purpose administrative areas.

Among the other special purpose authorities that are especially relevant to the rural sector, cooperative societies, rural development and Panchayati Raj deserve special mention. Cooperative societies are administered through a six-tier hierarchy: headquarters, division, district, sub-division, inspectors' circle and sub-inspectors' circle. Rural development and Panchyati Raj are five-tiered: headquarters, division, district, development block and village.

Environment management in Punjab, as in other states, remained neglected for long. It was only in 1980 that a separate Ministry of Environment was set up at the Centre. The state governments followed suit. Punjab got its first Department of Environment in 1981.

The Departments of Town & Country Planning, Public Works, Public Health Engineering, Soil Conservation and Housing and Urban Development were already looking after different aspects of the environment, but with very limited success. These authorities arrange their spatial administration in different ways. The Public Works Department (PWD) is organised in five tiers; Public Health Engineering, Soil Conservation, Water Supply and Sewerage Board and Housing and Urban Development in four; Town and Country Planning and Housing Board in three; and Water Pollution Board and Land Use Board in only one. Their boundaries generally differ from those of general purpose administrative areas.

The spatial organisation of the Regional Planning Department of Punjab deserves special mention. Punjab was earlier divided into three circles: Chandigarh, Amritsar and Hoshiarpur. The most striking feature of this arrangement was that each circle tended to enclose highly diverse physio-socio-economic areas. The clubbing of Ludhiana and Hoshiarpur districts in the same circle was one such example. Since 1995, the 17 districts of the state have got organised into five circles, each displaying considerable homogeneity within (Map 7).

Most of the state departments dealing with environmental services are either offshoots of some already existing department or of recent origin. Soil Conservation was earlier a part of the Department of Agriculture and Public Health Engineering of the Public Works Department. The Public Works Department itself was a part of the Railways Department during the early years of British rule. The spatial framework of administration in all such cases was inherited as a legacy. It displays some distortions if looked at purely from the viewpoint of requirement of a particular environmental service today.

Regional Authorities

These are constituted to manage and provide some higher order services specific to a particular function or area. Electricity boards, transport corporations and river valley authorities are some such arrangements. Generally, these require coordination among a number of general purpose authorities (East and Prescott, 1975).

On the basis of the spatial scale of their operations, regional level authorities may be classified into three categories: (i) national level, which treats Punjab as one of the spatial units of a countrywide arrangement operated by the Central government; (ii) state level, which covers the whole of Punjab and is a state responsibility; and (iii) sub-state level, which covers a part of the state and falls under the purview of the state government.

National level regional authorities include, among others, Indian Railways, Food Corporation of India, Fertiliser Corporation of India, Oil and Natural Gas Commission, Coal India, Steel Authority of India and Cement Corporation of India. Most of these treat Punjab as a part of their Northern Zone.

State-level regional authorities include, among others, Punjab State Electricity Board (PSEB), Punjab Roadways,

Maps 7 & 8

Punjab: Spatial Organisations for Administration

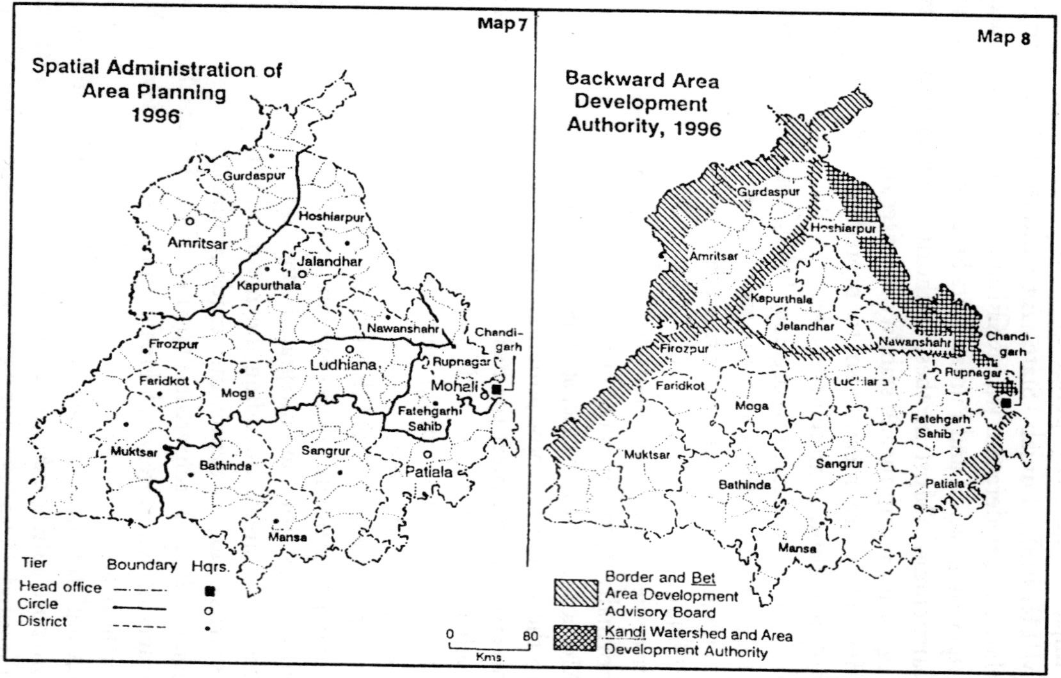

Punjab State Food and Civil Supplies Corporation (PUNSUP), Punjab State Co-operative Milk Producers' Federation (MILKFED), Punjab State Agriculture Co-operative and Marketing Federation (MARKFED), Punjab State Agro-Industrial Corporation and Punjab State Small Industries and Export Corporation. Most of these authorities are meant to ensure regularity in supply and stability of prices of different commodities. They are meant also to protect the interests of both producers and consumers.

These authorities organise their spatial administration along different lines, and the number of spatial tiers vary in their case. For example, PSEB organises its activities in six tiers; PUNSUP in four tiers; and MARKFED and MILKFED in three tiers each. The intensity of the spatial spread of a service normally determines the number of tiers for its administration. Moreover, in their case, spatial organisation may not conform to the general purpose administrative areas. It is determined in diverse ways to meet the technical and service requirements of each authority.

Sub-regional authorities are not function-based, but locality based, being specific to a particular area earmarked either for its integrated development, or for the solution of an endemic problem. The Punjab government has organised authorities for an integrated development of the state's backward areas, including the border districts of Gurdaspur, Amritsar and Firozpur; the *bet* (flood plain) areas of Ludhiana, Jalandhar and Kapurthala districts; and the *kandi* (foothill) tracts of Rupnagar and Hoshiarpur districts (Map 8). These authorities are vested with the powers essential for coordinating the activities of different general purpose and special purpose authorities. Under the state plans, additional funds are allocated for the development of these backward areas.

Conclusion

The spatial organisation of administration in Punjab since Independence represents in part the persistence of the colonial pattern and in part decolonisation. The basal structure of the administrative areas and the non-elective nature of the district administration remained practically the same. On the other hand, the change was represented by the creation of the development block as a new tier for rural development and a rapid increase in the number of special purpose and regional authorities to promote overall and regional development. On the whole, a shift in the underlying philosophy from law and order maintenance and revenue collection during the British days to development promotion during the post-Independence era is unmistakable.

In point of fact, one can discern three phases in the process: (i) the phase of administrative area reforms during 1947-66, partly as a byproduct of successive reorganisations and partly by way of formation of development blocks; (ii) the phase of proliferation of special purpose and regional authorities during 1966-85, with a view to diversifying the development activities and taking care of problem areas; and (iii) the phase of populism since 1985, which was marked by the creation of several new districts, sub-tehsils and development blocks in a short span of time.

In the process, the administrative organisation of space in Punjab has been assuming a more complex form over the years. Not only were new layers of administrative arrangement created, but the existing administrative units were also subdivided or readjusted in several ways. All this was not free from certain infirmities. For this, one requires a simple look at the administrative map of Punjab as it has taken shape after the formation of five new districts during the 1990s.

In fact, a number of issues need to be resolved before we arrive at the right kind of decisions. The first pertains to the determination of size and shape of the district. Three main guiding principles for this include maximisation of access of people to administration; effective supervision by the district administration, and minimisation of the cost to the system. A rule of thumb suggests that the farthest place in the district should not be more distant than an hour's bus journey from the headquarters. Such a distance could be reasonably placed at about 25 km. Accordingly, a district should cover an area of about 2,000 km^2. Punjab may have 25 districts as per this computation.

The second issue relates to the methodology for delimitation of districts. What kind of an area should a district enclose? Must it be marked by a commonality of local historical heritage or identity of people defined in some manner or similarity of physical geography? Should it meet the requirements of delineating meaningful legislative and parliamentary constituencies? Above all, to what extent should the existing administrative boundaries be honoured?

Finally, a vital issue concerns the number of tiers in which the administrative units of different hierarchical order should be organised. A pertinent question is: Should Punjab continue with the existing system of divisions, districts, sub-divisions, and development blocks or dispense with the divisions and sub-divisions and retain the districts and development blocks only? What is the relevance of a division in a small state like Punjab? Do we need sub-divisions when the number of development blocks in a district averages eight? Multiplicity of tiers in administration is always generative of confusion and delays.

References

Beaujeu-Garnier, J. (1976): *Methods and Perspectives in Geography*, Longman, London.

Bennett, R.J., and Chorley, R.J. (1978): *Environmental Systems – Philosophy, Analysis and Control*, Methuen, London.

Blazek, M. (1976): 'Administrative Regionalisation as a Problem of Geography' in V.C. Misra et al. (eds), *Essays in Applied Geography*, University of Saugar, Saugar.

East, W.G., and Prescott, J.R.V. (1975): *Our Fragmented World*, Macmillan, London.

Garner, B. (1975): The Effect of Local Government Reform on Access to Public Services: A Case Study of Denmark' in R. Peet et al. (eds), *Processes in Physical and Human Geography*, Heinemann, London.

Humes, S., and Martin, E.M. (1961): The Structure of Local Government Throughout the World, Martinus Nishoff, The Hague.

Kant, S. (1988): *Administrative Geography of India*, Rawat, Jaipur.

Krishan, G. (1988): 'The World Pattern of Administrative Area Reform', *Geographical Journal*, 154, 93-99.

Krishan, G. (1990): 'Administrative Geography', *Annals of the Association of Indian Geographers*, 10, 41-51.

Massam, B.H. (1972): 'The Spatial Structure of Administrative Systems', Association of American Geographers, Resource Paper No. 12.

Sharma, S.K. (1971): Deputy Commissioner in Punjab, Indian Institute of Public Administration, New Delhi.

Shukla, J.D. (1976): *State and District Administration in India*, National, New Delhi.

9

Development Problems of Rural Settlements in Difficult Ecological Environments: Cases from Gujarat

Anjana Desai

Introduction

The socioeconomic development of rural settlements depends on favourable physical environment (Adams, 1990). It is difficult to have sustainable primary activity in rural settlements if the physical conditions are hazardous. It has been noted that in many rural areas, land has been degraded over a period of time; water has not been harvested and forests, not protected. As a result, there has been depopulation in the rural areas, leading to out-migration, as rural sustainability to hold its population has decreased over time. In reality, we find that there is no employment in villages nearly nine months a year. Only one crop is harvested, which is not enough to sustain the people (Chen Martha, 1991). Resources are not managed, upgraded and conserved on a sustainable basis. Non-farm activities are not introduced or developed along with farm activities.

The rural population is caught in a vicious circle of rural-urban migration on the one hand, and land degradation on the other (Lahiri, 1980). The objectives of this paper are to find out (a) the extent to which physical parameters are responsible for the socioeconomic and demographic growth of settlements in three physically weak ecosystems in Gujarat: the Bhal region – a coastal, marine ingressed area; the Nal region – a saline lacustrine area; and the Panchal region – a hilly, eroded and drought prone area; and (b) the extent to which alternative non-farm economic activities for the economic sustainability of the people are feasible in these regions.

Methodology

Physical parameters have been used to classify the three regions by land capability, using GIS. Various physical parameters have been superimposed with the help of GIS to know the composite potentiality of a piece of land (Davidson, 1992). The negative impact of land degradation on the growth of settlements has been measured in terms of changes in population density, population potential, size of settlements, land use, cropped area, crop yield, levels of sustainability per unit area and per person over time using primary and secondary sources of data, viz. remote sensing data from the Space Applications Centre, agricultural data from the Department of Agriculture, government of Gujarat and population data from the Census.

Study Area

The three backward regions, Bhal, Nal and Panchal, occupy an axis extending from Northwest to Southeast, which separated Saurashtra geologically from the mainland of Gujarat during the Pleistocene-Pliocene period. Later, during a more recent period, the region was lifted up and filled up with riverine and marine deposits. The backwardness of these regions is revealed by their low quality of life in terms of earnings, consumption, assets, the socioeconomic and technological background of the people in sustaining themselves on land, which suffers from various types of physical hazards.

The study area covers three fragile ecosystems. The Bhal region, covering 275 villages and occupying an area of 4,392 km^2 in the districts of Bhavnagar, Surendranagar, Ahmedabad and Kheda, is a marine ingressed, flood affected, coastal region surrounding the Gulf of Khambhat. The Nal region, lying to the North of the Bhal region in Ahmedabad and Surendranagar districts, with 81 villages and spread over 1,622 km^2, is a salinity prone, lacustrine basin of the Nal lake. Further Northwest is the Panchal region, which is a hilly and drought prone area, with 346 settlements, three urban centres and a total area of 3,170 km^2, lying in the districts of Rajkot, Surendranagar and Amreli (Figure 1).

Figure 1
Gujarat: Location of Study Area

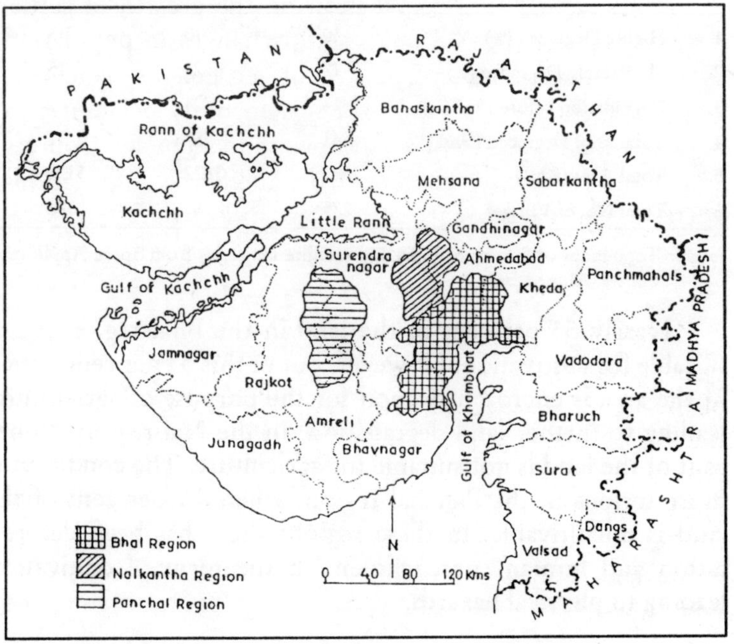

Analytical Results

The main problem relating to the growth of rural settlements in the area is degradation of cultivable land into wasteland. Nearly 57 per cent of the land in the Nal region, 41 per cent in the Bhal region and 32 per cent in the Panchal region is degraded (Table 1). The reasons of degradation are both physical hazards and human interference and negligence. In the Bhal region, marine ingression in the coastal region has increased soil and water salinity. Irrigation with saline water and deforestation of coastal mangroves leading to marine ingression represent human mismanagement because of which many villages have become unproductive wasteland; the natural consequence is out-migration.

Table 1
Degradation of Land (1967-88)

	Levels of Degradation	*Bhalkantha*	*Nalkantha*	*Panchal*
1.	Highly Degraded (%)	10	10	15
2.	Moderately Degraded (%)	17	23	15
3.	Degradation before 1967 (%)	14	24	2
4.	Total % of Degraded Land (%)	41	57	32
5.	Total Area (Km^2)	4392	1622	3170
6.	Total No. of Villages	275	81	346

Source: Topo-maps of Survey of India and satellite imageries from Space Application Centre, Ahmedabad.

Nearly 53 per cent of the land in the Bhal region is not suitable for cultivation. However, out of this 53 per cent, much of the area is encroached upon for the purpose of agriculture, leading to further land degradation. In the Nal region, 75 per cent of the land is not suitable for agriculture. The condition is more serious in the Panchal region, where 79 per cent of the land is uncultivable. In these regions there has been deforestation and human encroachment in the form of cultivation leading to physical hazards.

Table 2

Change in Density of Population (in km^2) during 1971-91

Bhalkantha Region	*No. of Villages*
1. High (41 and Above)	82
2. Medium (31-40)	32
3. Moderate (21-30)	28
4. Low (11-20)	28
5. Very Low (0-10)	28
6. Negative Change	29
Total	227
Nalkantha Region	
1. High (141-175)	1
2. Medium (106-140)	1
3. Moderate (71-105)	3
4. Low (36-70)	19
5. Very Low (1.35)	52
6. Negative Change	2
7. Uninhabited	3
Total	81
Panchal Region	
1. High (160 and Above)	5
2. Medium (121-160)	4
3. Moderate (81-120)	30
4. Low (41-80)	102
5. Very Low (1-40)	189
6. Negative Change	16
Total	346

Source: Census of India

The impact of land degradation is clearly seen on the demographic profile of the settlements in terms of density, size distribution and population potential during 1971-91 (Tables 2, 3 and 4). In the Bhal region, out of 227 villages, 29 villages had a negative change in density. The figure is 16 for the Panchal region and two for the Nalkantha region. In case of size distribution of settlements, the smaller ones have lost population and the medium ones have gained it during 1971-91 (Table 3). In terms of size distribution, upgraded settlements are more in the Panchal region (61%) followed by the Bhal (42%) and the Nal (38%). Similarly, settlements which remained in the same class are more in number in the Nal region (58%), followed by the Bhal (50%) and the Panchal region (39%). Growth of settlements is better in the Panchal region due to the presence of mineral resources like fire clay, silica, sandstone and limestone near the Chotila hills and in the Thangarh area. These are useful in developing small scale industries of pottery and sanitaryware as an employment opportunity for the local people. As a result, there is a lower rate of outmigration and the change of density is higher near the zone of mineral resources, where the percentage of non-primary workers varies between 15 per cent and 80 per cent (1991). The presence of non-primary workers is low in the Nal region. Four villages with a higher percentage of non-workers are seen in the relatively prosperous parts of the Dhanduka and Dhoka talukas of the Bhal region.

Table 3
Change in Distribution of Size of Settlements

Variables	*Bhalkantha*		*Nalkantha*		*Panchal*	
	1971	*1991*	*1971*	*1991*	*1971*	*1991*
Less than 200	21	14	–	–	46	10
200-499	52	36	15	6	122	78
500-999	102	79	18	19	117	123
1,000-1,999	63	92	31	32	51	102
2,000-2,999	25	32	10	10	5	22
3,000-3,999	7	13	4	8	2	5
4,000-8,999	5	9	–	3	2	3
More than 9,000	–	–	–	–	1	3
Uninhabited	–	–	3	3	–	–
Total	275		81		346	

Source: Census of India, 1971, 1991.

The development of each settlement depends not only on the availability and utilisation of resources within the villages, but also on the nature and development of the surrounding settlements, which is reflected in the population potential of the settlements (Table 4).

Table 4
Population Potential

Population Potential in '000	*No. of Villages*	
	1971	*1991*
Bhal Region		
1. High (11-130)	-	43
2. Medium (91-110)	12	74
3. Moderate (71-90)	125	101
4. Low (51-70)	96	42
5. Very Low (31-50)	42	15
Nalkantha Region		
1. High (11-130)	-	3
2. Medium (91-110)	1	8
3. Moderate (71-90)	49	63
4. Low (51-70)	28	4
5. Very Low (31-50)	3	3
Panchal Region		
1. High (11-130)	-	3
2. Medium (91-110)	-	53
3. Moderate (71-90)	3	201
4. Low (51-70)	202	88
5. Very Low (31-50)	141	1

Population Potential $P_i = M_i$
Where M_j = Population of j and
Dij = distance between I and j (It is half distance to the nearest neighbour, when I – j)

High population potential is expected when settlements are large in size and closely spaced, indicating a high level of growth and interaction in terms of trade and business. The population potential has changed over time (1971-91) and space in these regions. In 1971, there was hardly any settlement with high and medium potential, while in 1991, some settlements showed a high potential in all three regions. In the Bhal region, high population potential is seen in the interior areas of Kheda, Surendranagar and Ahmedabad districts, while in the Nal region, high potential is seen in the Northern section in Ahmedabad district. In the Panchal region, three urban centres, Thangarh, Chotila and Vinchiya, in the Central zone, had high population potential in 1991. The peripheral zone has a lower potential.

The socioeconomic development of the settlements is much dependent on the social infrastructure, even if resources are not locally available. If a region is well connected by roads and endowed with power supply, it can develop its own industries based on external raw materials. Nearly 86 per cent of the total settlements had a very low level of amenities in the Bhal region in 1971. It was worse in the Nal and the Panchal regions, where 95 per cent of all settlements had very low levels of amenities. In all the regions, there was an improvement in the availability of basic amenities in 1991.

Development of settlements depends on the economic sustainability of a region, dealing with aspects of employment, per capita income, man-days of work, wage rate, savings, assets, level of consumption and ability to spend. It is seen that these areas, by any standard, are not economically sustainable. The backwardness is related to lack of information on land capability, alternative occupation, appropriate technology, procurement of finance, marketing and general organisational aspects. It is noticed that wherever farmers are organised through cooperatives, NGOs or VLWs, they have prospered; otherwise, in general, there has been a negative change in the cropped area and yield rate.

The development potential of the settlements has been mapped for 1991 by covering a large number of weighted parameters – physical, demographic and socioeconomic. More than 50 per cent of the villages of the Bhal region fall in the category of low or very low potential for development. These villages are located mainly in the coastal belt, where water-logging and soil salinity cause major hindrances to development. The coastal settlements have experienced depopulation through the process of out-migration due to economic and environmental unsustainability. The condition is worse in the Nal region, where 62 per cent of the settlements have low and very low potential for development. Relatively, fair potential is found in the Eastern area, where the land capability is better. In the Panchal region, nearly 73 per cent of all settlements exhibit low and very low development potential.

It is thus clear that the developmental potential of the settlements in all these regions is very low and they lack in individual and organisational management in terms of training, alternative economic opportunities, production and marketing.

In order to know the problems and prospects of development of settlements in difficult ecosystems, it is important to classify land in relation to its capability based on physical parameters. Then land use can be compared with the proposed land use in order to obtain better management of the land. The suitability of land for agriculture is grouped in classes I to IV, while mixed farming like dairy, fishing, agro-based or mineral based industries, mushroom or sericulture, forestry and other non-farm activities can be combined with agricultural activities on land, which is grouped in classes V to VIII. It is quite possible that all types of land are not suitable for agriculture. In fact, land degradation has taken place wherever an encroachment has been made by agricultural activity in areas that are not suitable for it. In order to check land degradation by human interference, it is important to use land as per its capability based on major physical conditions like relief, climate, soil, quality of groundwater, vegetation cover, etc.

The Bhal region can be divided into three sub-regions, based on land capability: coastal, sub-coastal and inland. At present the coastal area is characterised by saline, marshy mud-flats in groups of land capability of VII and VIII, with very low density of population. The proposed land use is fishing, salt manufacturing, growing of salt tolerant grass and trees. The sub-coastal area is now fallow land in classes V and VI, which can be used for pasture and cultivation of coarse grain. Agriculture can be introduced with water harvesting and soil conservation.

The Nalkantha region has seven sub-divisions as per level of land capability. The Northeastern area belongs to classes II to IV. Hence, intensive cultivation is proposed, instead of extensive. The Western part, in classes V to VIII, has eroded and barren land which needs conservation of soil through plantation of trees, pasture and cultivation of millets. The central area, around the Nal lake, is highly saline and marshy in nature. The proposed land uses are fishing, recreation and plantation of salt-resistant trees.

The Panchal region can be grouped into three sub-regions based on land capability. The Northern part belongs to classes III and IV, where cultivation can be improved with water harvesting and soil conservation. The land around the central Chotila hill, belongs to the categories V and VI, where cultivation can be practised with grassland, which can be utilised for the rearing of livestock. Cultivation needs to be avoided on the slopes so that the land does not get exposed to erosion and run-off. The central hilly area, used for cultivation and fallow land, which is in class VIII, is proposed to be used mainly for recreation and plantation of trees to check soil erosion.

Problems and Prospect of Rural Sustainability

It is important to note that all these regions suffer from both physical hazards and poor management of land resources. There has been increasing land degradation, leading to loss of agricultural land and production per acre and per person. Instead of conserving soil and water, farmers have encroached upon areas for cultivation which were meant for pasture and forests. As a result, more land has been exposed to soil erosion and degradation and, with more out-migration of labour from these lands, left without upgradation or maintenance.

A proper long-term management of land resources, as seen in China and Japan, can bring Indian farmers in similar environments out of poverty, providing sustainable employment, earnings, amenities and better lifestyles, which would be useful in decentralising the developmental processes. The following steps in micro-level planning can be taken up, depending on the needs of the eco-system:

(a) Land-use planning to check soil erosion and conservation of soil moisture with suitable grass and forest cover in areas that are not suitable for agriculture;
(b) Introduction of new techniques of water harvesting through sprinkle and drip irrigation, check dams and collection wells for rainwater harvesting;
(c) A new cropping pattern bringing in crops which grow well with less water. Vegetable and fruits, such as beetroot or berries resistant to salinity, need to be grown;
(d) Introduction of better inputs like hybrid seeds, organic manure and new farming techniques to increase yield;
(e) Initiation of other farm related ancillary activities to make farmers work on a cooperative basis on land that is unsuitable for agriculture; for example, dairy and poultry farming, fishing, mushroom growing, sericulture, cotton ginning;
(f) Introduction of non-farm activities for diversification of occupation as a check against physical hazards. Crops

often fail due to drought and flood. Farmers need to have alternative occupations for their survival. Various types of small-scale industries, like food processing, silk industry, cotton ginning, brick making, etc., would certainly increase the purchasing power of farmers with full employment in the region itself;

(g) Reclamation and maintenance of land productivity on the one hand, and improvement of soil on the other, for making the farmers self-sufficient; and

(h) Provision of government subsidy and fair pricing system, availability of information on farm techniques and vocational training centre, marketing assistance for farm and non-farm products as parts of the holistic programme of rural sustainability.

The fundamental need is to increase opportunities for employment in villages so that the farmers are not required to move out of their areas. Such a plan would be able to reduce the socioeconomic gap between 'people' and 'regions'. Only an increase in total food production does not guarantee its proper distribution. In reality, it is skewed due to the unequal distribution of purchasing power. Hence, it is all the more essential to develop weak regions with special programmes. Such a developmental plan would ensure improved earnings and higher consumption of nutritious food by farmers. Finally, self-sustained villages with amenities will not only check rural-urban migration, but also attract people from urban areas to rural regions that possess a peaceful and clean environment for living and working.

Acknowledgement

The findings of this paper are based on the project, 'Management of Land Resources through Geographic Information System', funded by the Department of Science and Technology, Government of India, New Delhi, during 1992-95.

References

Adams, W.M. (1990): *Green Development, Environment and Sustainability in the Third World*, Routledge, London.

Chen Martha, A. (1991): *Coping with Seasonality and Drought*, Sage Publication, New Delhi.

Davidson, D.A. (1992): *The Evaluation of Land Resources*, Longman Scientific and Technical, England.

Lahiri, T.B. (1980): *Rural Development*, Jayashree Prakashan, Kolkata.

10

Fluvial Environment and Development of the North Bengal Plains with Special Reference to Land-use Problems

S.C. Mukhopadhyay

Introduction

This study relates to the fluvial environment and development of the North Bengal Plains with special reference to land-use problems. It also makes an attempt to investigate the evolutionary episodes of the drainage basins of the Eastern Himalayas, such as the Mahananda, Mechi, Tista, Jaldhaka, Torsa, etc., and the diverse kinds of resource potential. Modern methodology (i.e. pre-field, field and post-field methods) and intensive field investigations in some selected parts of the Mahananda, Tista, Jaldhaka, Torsa and adjacent drainage basins in the Eastern Himalayas offer ample scope for studying the multi-disciplinary complexities of fluvial geomorphology under varied structural and climatic conditions. The behaviour of the Eastern Himalayan river system relating to channel dynamics by using the Turkey Brook bed-load transport data

(Mukhopadhyay, 1982) is primarily dependent on the interaction of geology, pedogeomorphology and hydraulics on macro-meso-micro scales. Tectonics and lithology, being important components of geology and geomorphology, have a profound influence on the behaviour of these mighty river channels primarily within the Brahmaputra drainage basin. Here, a variety of drainage and channel patterns have resulted from and reflect on different combinations of hydrological conditions, sediment characteristics and the geological history of these rivers and drainage basins.

Physiographically, the North Bengal Plains are divided into three distinct parts: (i) the northern part, occupied by low ranges of the Himalayan foothills (Pre-Quaternary) and traversed by a terraced piedmont zone (300-1,000 m); (ii) the central part made of vast alluvial, pebbly, riverine flats set in a series of conjugate-alluvial fans (Pleistocene) and drained by major rivers like the Tista, Jaldhaka, Torsa (the Brahmaputra Drainage), Mahananda, Balasan and their innumerable tributaries (the Ganga Drainage), (200-300 m); and (iii) the Southern marginal or lower plains of terraces and flood plains with alternate meander belts and subdued interfluves (100-200 m). The general slope of the terrain is from Northwest to Southeast and is intersected by rivers flowing generally from North to South in response to the underlying structure and tectonics.

Assemblages of peculiar land forms are found to be developed primarily due to the occurrence of varied sculpturing processes. These are mostly of fluvial and tectonic origin and are categorised into three main groups: (i) valley-fill deposits along the major streams within the hilly tract, forming alluvial and aggradational terraces; (ii) older fan deposits forming terraces (cut and fill type) along the piedmont zone; and (iii) flood plain deposits and recent fans in the plains. On the basis of breaks in slope, variations in the constituent materials, degree of consolidation, index of oxidation, effect of erosion and land-use practices, several geomorphic units containing 15 pedogeomorphic surfaces have been identified in the area under consideration (Mukhopadhyay, 1982; 1994; 1995; Majumdar et al., 1991; Das et al., 1993). The prevailing geomorphic processes owe their origin largely to climatic conditions.

Broadly speaking, the area falls under the modified tropical monsoon climatic regime and the overall climate can be grouped seasonwise as: (i) cold weather season – December to February; (ii) hot weather season – March to May; (iii) Southwest monsoon season – June to September; and (iv) retreating monsoon – October to November.

The northern part of the study area experiences the humid sub-tropical type of climate. The mean average temperature in the coldest months is 5°C (January) and that in the monsoon months, 22°C (July). Average annual rainfall exceeds 3,500 mm. On the other hand, the southern part of the plain area is characterised by mean temperatures of 18°C in winter and 30°C in summer, which has exerted a strong influence on the development of land-use, especially the agricultural land-use patterns of the area.

Major Drainage Characteristics

The drainage system of the North Bengal Plain is typified by the occurrence of complex-compound-composite peculiarities. These drainage channels can be broadly classified into four river basins. From West to East, they are (i) the Mahananda river basin; (ii) the Tista river basin; (iii) the Jaldhaka river basin; and (iv) the Torsa-Raidak-Sankosh river basin. It seems reasonable to discuss some relevant points of the upper reaches of these river valleys. Of them, the Mahananda is a sub-basin of the Ganga river system, meeting the Ganga trunk channel to the upstream of Farakka. The other three are the sub-basins of the Bramhaputra river system. The Tista river, which is the largest snowfed perennial river in the area, used to flow into the Ganga river system through the now moribund channel of the Karotoya river about two centuries ago. At least three successive Eastward migrating channels of the Tista can be surfacially recognised between the present-day Karotoya and

Tista river channels in the Basubri area (26°24'N, 88°44'E). Actually, the Mahananda and its tributaries on the Western border originate in the Lesser Himalayas and the foothills, whereas the Tista originates in the Pauhunri glacier of the Greater Himalayas. In the course of its journey to the plains, it collects the flow of a number of tributaries originating from the Greater and Lesser Himalayas and also from the foothills. The Jaldhaka, Torsa and Sankosh rivers originate in the lower reaches of the Greater Himalayas and receive the flow of a number of tributaries and sub-tributaries coming down from the Lesser Himalayas and the foothills. The Raidak and the Kaljani rivers also originate in the Lesser Himalayas. Some aspects of fluvial environment, soil conditions, rainfall, water discharge, erosion status, etc., are shown in Table 1 in terms of five stations corresponding to the pedogeomorphic surfaces in the North Bengal Plains. The snow-melt water makes a substantial contribution to the Tista and a limited contribution to the Jaldhaka and Torsa river regimes. The rest of the rivers are essentially groundwater fed. This is clearly reflected in the quantum of base flow and the magnitude of fluctuations in their seasonal discharge. The drainage configurations in the upper catchment of the river basins in the Eastern Himalayas have evolved through a combined control of structural fabric and lithology.

The lithology has an important role over structures in shaping the drainage net upto the 32nd order (Strahler Number) channel segments. Structure plays a predominant role in the alignment of the higher order channels. In the alluvial plains, the configurations of the channels have been controlled by the neo-tectonic fabric of the area, aided by the process of alluviation. The channel avulsions have resulted from the combined effects of landscape degradation, dynamic slope and fluvial processes and local earthquakes. The annual floods are normally influenced by climate and fluvial processes and have a predictable pattern of inundation in a given channel network.

Table 1

Fluvial Environment, Discharges and Erosion Status of Five Pedogeosurfaces

Pedogeosurface (Morphoattribute) Stations (Altitude in Metres, a.s.l.)	*Rainfall (mm) Discharge (m^3 sec^{-1}) in P.M., M. & P.O. (P.M. Pre-monsoon, P.O. Post monsoon, M: Monsoon)*	*Erosion Status Soil Loss and Run-off (in m^3 $sec^{-1}km^2$) and Geosensitivity (G)*	*Remarks (Explanation)*
Dalsinghpara (270 m)	P.M. 810: 3600 m^3sec^{-1}	Eh R 270	Rainfall (mm)
a =40, b=15'	M 3820: 8200 m^3sec^{-1}	SL 12-25	Discharge (m^3 sec^{-1}) for Pre-monsoon (P.M.),
c=11, d=25	P.O. 23: 4400 m^3sec^{-1}	RF 12-16, GI	Monsoon (M) and Post-Monsoon (P.O.) Seasons,
Kalchini (225 m)	P.M. 640: 1600 m^3sec^{-1}	Eh R 180	RF=Runoff (m^3 $sec-^1km^{-2}$).
A=25, b=10°	M 3200: 3800 m^3sec^{-1}	SL 6-12	Morpho-attributes, a: Rel.Relief (m) b: Slope-AV
c=18, d=20	P.O. 22: 2000 m^3sec^{-1}	RF 10-14; GI	(degrees)
Hasimara (220 m)	P.M. 810: 3800 m^3sec^{-1}	Eh R 260	c: Drainage Density km./sq.km.
a=20, b=8°	M 3765: 8000 m^3sec^{-1}	SL 10-20	d: Roughness Index
c=90, d=18	P.O. 21: 4900 m^3sec^{-1}	RF 11-18, GH	e-Erosion Status (Erosion Hazard Assessment on Erosivity Values)
Matiali (220 m)	P.M. 730: 1200 m^3sec^{-1}	Eh R 450	h: High Erosion Risk,
a=200, b=8°	M 3685: 3400 m^3sec^{-1}	SL 12-20	m: Moderate, etc. (Stocking and Elwel (1976)
c=15, d=55	P.O. 19: 1800 m^3sec^{-1}	RF 16-26 GH	
Sevok (260 m)	P.M. 480: 120 m^3sec^{-1}	Eh R 300	
A=160, b=15°	M 2820: 3100 m^3sec^{-1}	SL 21-36	
c=10, d=58	P.O. 15: 2700 m^3sec^{-1}	RF 20-38, Gvh	

Source: Indian Met. Deptt, C.W.C., Computed and estimated (Mukhopadhyay 1982, 1989, 1995, 1996)
R-Rainfall Erosion Index, El 30/140, Erosion in Foot-tons/Acre, I_{30} Inches/Hour (Wischmeier and Smith 1978), SL: Soil Loss, Tonnes/Ha/Year (Estimated). Morgan (1974), G : Geomorphological Sensitivity, GH. High GI. Intermediate, etc.

The major rivers which originate from the Himalayas are mostly antecedent (pre-Quaternary) drainage channels and have been responsible for the alluviation of the foredeep belt. The consequent drainage net is a neo-drainage system (Figures 1 & 2). As mentioned earlier, the important rivers of this area originate from the snow-clad Himalayan hills and glaciers. After cutting across the ridges and mountain ranges, they debouch into the plains of North Bengal, carrying huge loads of sediments both in suspension and traction. Most of the bedload in the form of coarse clastics are deposited in the foothill areas as fan deposits. The coalescence of such fan deposits of the various rivers, flowing Southward, give rise to the vast piedmont zone. It is interesting to note that in the whole Himalayan foothill zone, the piedmont zone of the Kosi-Torsa belt is the widest. This may be attributed either to heavy precipitation in a short span of time or rapid upheaval of the hill ranges forming the catchment of these rivers or to both factors acting together. The following sections will highlight the trend of behaviour of all such river channels originating in this vast catchment zone. In a general way, the drainage pattern of the major rivers in the area is anastomosing or of the braided type due to the heavy load of sediments on the river bed. On the other hand, within the hilly tracts, the sinuous channel courses, parallel to sub-parallel, rectangular, trellis type and pinnate patterns, are all indicative of the underlying structural control.

The major tributaries of the Tista are the Gish, Lish, Chel, Neora, Dharla, etc., while those of the Jaldhaka are the Kurti, Murti, Diana and Dudhuya. Holong, Bania, Paro, Kurai, Thuria, etc., form the tributaries to the Torsa. With high levels of rainfall, influx of huge loads of sediments, steep gradient and vast alluvial plains, The South Sikkim, Bhutan and North Bengal rivers are unique in their behaviour as well as environmental changes. For example, the river Tista with all its tributaries, has a converging system where it meets with the great Rangit river, south of Rangpo. South of this confluence point, there is a diverging system (Sevak) of the river basin,

Figure 1
General Geological Map of North Bengal Plain, W.B.

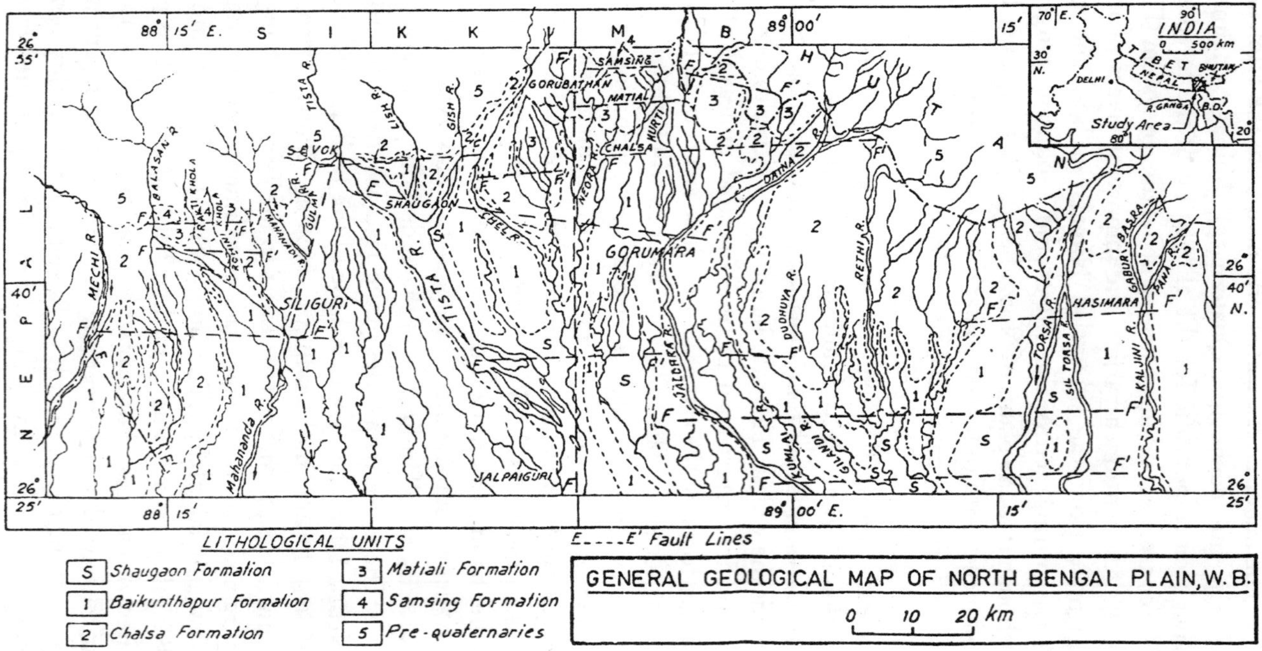

Figure 2
Quaternary Geological Map of Tista-Jaldhaka-Torsa-Mahananda Basin, West Bengal

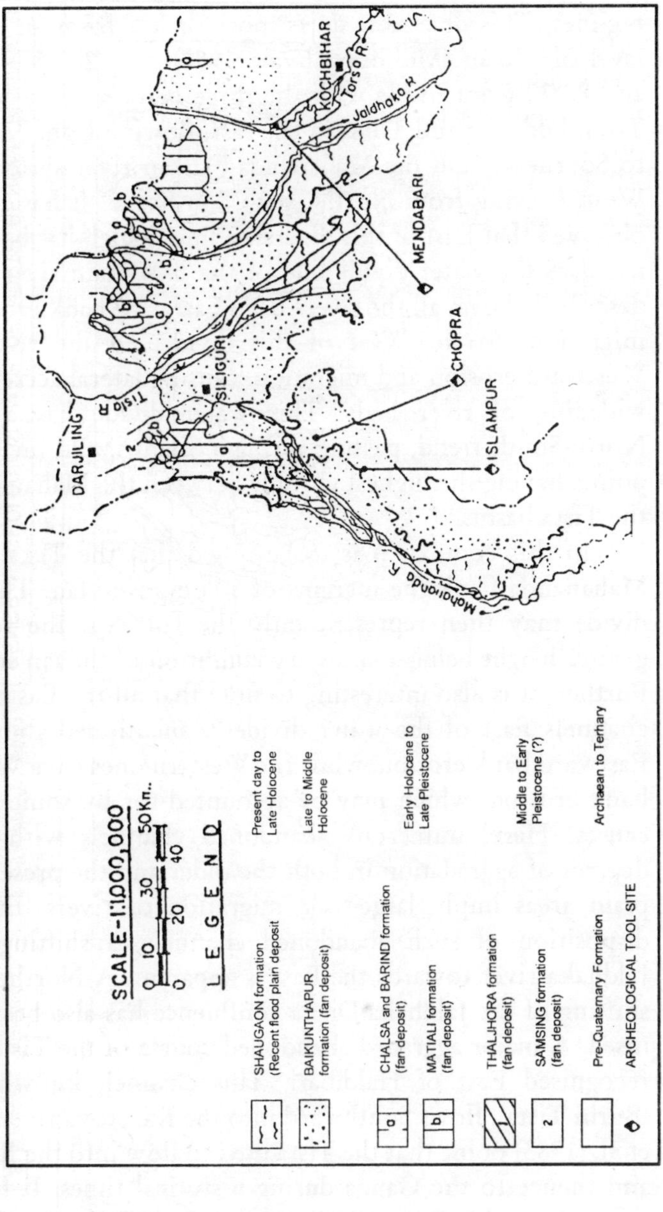

developing numerous fans in the area in successive flood regimes. This diversion starts more or less from an elevation level of 300 m (Mukhopadhyay, 1982) near 26°45'N latitude and 88°30'E longitude with the three main river systems, the Tista, Jaldhaka and Torsa in the east, flowing from Northwest to Southeast, and the Mahananda-Balasan river system in the West, flowing from Northeast to Southwest. It has also been observed that East of this diversion line, which has been designated as the water divide line of the two main river systems described above, all the rivers have Eastward bank erosion and migration; to the West of this divide, all the rivers have Westward erosion and migration, causing lateral accretion and widening of river beds. This water divide has a general North-South trend, passing through Siliguri, and includes the points having the highest altitude between the Mahananda and the Tista basins.

In this context, it is to be noted that the Tista and the Mahananda flank the margins of a composite fan. This water divide may then represent only the fan axis, the relatively greater height being a necessary condition of the fan structure. Further, it is also interesting to note that all the Eastern river channels East of the water divide as mentioned above, have Eastward bank erosion, while the Western ones have Westward bank erosion, which may be accounted for by some tectonic causes. Here, numerous abandoned channels with various degrees of aggradation in both the older and the present flood plain areas imply large-scale migration of rivers. From the disposition of such abandoned channels, a shifting of the Jaldhaka river towards the East is apparent. A Northeastward shifting of the Jaldhaka-Diana confluence has also been recognised. A major aggraded abandoned course of the Tista can be recognised East of Haldibari. This channel, known as the 'Burhi Tista', flows Southward into the Karotoya river. Bagchi et al, (1983) opine that the Tista used to flow into the Karotoya and thence to the Ganga during historical times. It has been pointed out by Das and Chattopadyay (1993) that the river

Tista rising in the Sikkim mountains, when surveyed by Major Rennell (1779), took a course due South after running past Jalpaiguri, joined the Attreyee and, after flowing past Dinajpur, joined the Natore river and had a confluence with the Ganga at Jafferganj. In 1787, either one or two years after Major Rennel's survey was completed, an unusual flood occurred; the river brought down from the hills sufficient quantity of twigs and stones to throw a dam across its junction with the Attreyee and, taking a Southeastward course, it joined the Brahmaputra above Dewanganj in Bangladesh. The river Torsa also has shifted Eastwards from its original course, South of Hasimara. The general pattern of such shifting of channel courses may either be due to: (i) sudden high discharge during floods with concomitant capturing of earlier channels; (ii) neotectonic activities, causing uplift of some blocks of land, which resulted in consequent drainage channels; and (iii) the combined effect of (i) and (ii). The turning of North-South flowing rivers in a Northwest-Southeast direction like the river Tista near Sevok, Chel Nadi near Apalchand, Indog river near Gorumara, etc., indicate the presence of structural weak planes which have guided the courses of these rivers (Bagchi et al. 1983; Mukhopadhyay, 1989; 1991; 1994).

It is important to note that the Survey of India has run a number of high precision levelling lines across the Indo-Gangetic Plain and repeated them over the last 100 years or so. Analysis of these data has suggested vertical movement patterns along certain lines. They have concluded that there is a general subsidence in the Western sector, ranging from 1 mm per year to 31.5 mm per year, and uplift in the Eastern sector, ranging from 1 mm per year to 11 mm per year. It has been inferred that the tilt is associated with tectonic processes (Mukhopadhyay, 1996).

General Land-use and Eco-sustainable Development

The classification of the area into some morphogenetically significant terrain units (Table 1) has helped in identifying and delineating areas in terms of their susceptibility to flood hazard from the present-day river system. The evaluation of these terrain units in terms of other requirements to qualify as natural development corridors for various human activities has briefly brought out their potential for each such purpose.

(a) **Surface Communication Routes:** Both the Chalsa and Baikunthapur units provide the best grounds. The Matiali terrace is unsuitable for East-West roads across the drainage due to deep dissection, but it provides good ground for N-S feeder roads upto and into the hills. Hazard protection measures would be necessary elsewhere.

(b) **Urban and Industrial Complexes:** The Chalsa terrace areas in the Southern part of the piedmont plain and its flood plain equivalent to Baikunthapur terrace provides the best eco-sustainable development corridors for such purposes. In North Bengal, the terrain conditions do not permit the development of a metropolitan city. A medium sized urban complex, sustaining upto half a million population in a single unit, appears to be most suitable. Industrial complexes can act as satellites to such an urban area. Small urban industrial centres can be located in the southern part of the Matiali terrace area, where water would be available throughout the year.

(c) **Forestry:** The piedmont plain of North Bengal is best known for its natural forest wealth. But plantations would be eco-sustainable in the Samsing, Matiali, Chalsa and Baikunthapur units. The other two units should be left for forests. Social forestry has to be the selective activity complementary to cultivation, orchards and agro-plantations.

(d) **Agro-plantation:** The Matiali terrace is the best area for eco-sustainable development of tea plantations and orchards. The next best is the Chalsa terrace. The Samsing unit occupies third place. However, Chalsa and the higher areas of Baikunthapur are suitable for pineapple, guava, banana and jackfruit. The Samsing terrace is also good for orange orchards.

(e) **Kharif and Rabi Cultivation:** The Shaugaon plain and Baikunthapur terrace areas to the South of latitude 26°40'N are most suitable for such activities. In the area to the North, such activities become restricted and can be the secondary mode of economic sustenance of the local population.

Land-use Planning

With reference to the aforesaid discussions, it should be noted that the planners from different agencies will have to keep in mind the behaviour of the rivers, its causes and effects, so that the land can be properly used and water suitably utilised for common beneficial purposes.

1. At present, serious soil erosion from the hill slopes and high level terraces adjacent to the steep river banks are common in the entire Eastern Himalayan terrain and adjacent plains, especially in the North Bengal Plains. This soil erosion further triggers debris slides in the area. Proper reclamation of badland areas could be practised to avoid soil erosion to a considerable extent.
2. Occurrence of the shifting of rivers and serious bank migration and bank erosion problems Eastwards in the Tista-Torsa river system and Westwards in the Mahananda-Balasan river system, due to probable tectonic causes, should be taken seriously in consideration for any future development activities. The rivers changing their courses due to some structural weak planes in localised areas should also be taken into account for these purposes.
3. Large-scale bank erosion and bank failure have been observed to get aggravated during the post-monsoon period when the rivers become affluent and groundwater feeds the river by scooping out the unconsolidated sandy matrix. Proper deep-seated spurring with iron netting may be formulated to avoid all such hazards.

4. The detailed delineation of each of the terrain units, such as the Shaugaon plains, the Baikunthapur terrace, the Chalsa terrace, the Matiali terrace and the Samsing terrace as above (Mukhopadhyay, 1991; 1994; 1995), is required for eco-sustainable development planning; this can be accomplished rapidly and at a low cost by conjunctive application of remote sensing and selective ground verification techniques. There is a close relationship between the terrain units and land-use patterns, especially the development of plantations in the Baikunthapur (Southern part), Chalsa and Matiali units.

Concluding Remarks

The North Bengal Plain areas manifest the effects of variations of physico-chemical processes as well as the changing environmental and orogenic evolutionary relationships particularly those associated with the occurrence of alternating glacial-interglacial regions. The continuing neotectonic activities, for example, have been confirmed through geomorphological evidence as recorded in the upper Shiwalik structural features; post-Quaternary movements should also be considered in deciding on the developmental aspects of special land-use balance. All the geo-ecological factors discussed so far, affecting especially river behaviour and attitude, should be seriously considered for achieving any beneficial activity towards regional development of the densely populated areas of this part of Eastern India.

Acknowledgement

The author is grateful to Professor K. Bagchi, retired head, Department of Geography, Calcutta University; Professor A.K. Saha, former Emeritus Professor of Geology, Presidency College and secretary of the Centre for Study of Man and Environment (C.S.M.E.), Kolkata; and Dr D.R. Nandy, retired director of the G.S.I., for their kind advice and cooperation.

References

Bagchi, K., and Mukherjee, K.N. (1983): *Diagnostic Survey of West Bengal (North)*, Department of Geography, University of Calcutta, 37-46.

Das, A., and Chattopadhyay, G.S. (1993): 'Neotectonics in Tista, Jaldhaka and Torsa Interfluve Belt of North Bengal', Record, GIS, 121 (2-8), 101-109.

Majumdar, S., Sarkar, K., Basu, A.K., Sinha Roy, S.K., and Chakraborty, D.K. (1991): 'Report on the Study of Neotectonics and Shifting of Rivers vis-à-vis Erosion and Flood Incidence in Parts of the Himalayan Foredeep through Integration of Remote Sensed Data with Other Collateral Data', GSI Unpublished Report, F.S. 1989-91.

Mukhopadhyay, S.C. (1982): *The Tista Basin – A Study in Fluvial Geomorphology*, K.P. Bagchi and Co., Kolkata, New Delhi, 97-216.

— (1989): 'Geomorphology of the Ranikhola Basin Sikkim', *Indian Journal of Landscape Systems and Ecological Studies*, 12(2), 1-8.

— (1991): 'The Complex Structure and Geomorphological Reality of the Central Himalaya and its Surroundings', *Indian Journal of Landscape Systems and Ecological Studies*, 14(2), 128-43.

— (1994): 'Basinal Characteristics of the Middle Torsa Basin', *Indian Journal of Landscape Systems and Ecological Studies*, 17(2), 105-20.

— (1995): 'Fluvial Systems of the North and Eastern India Subcontinent with Special Reference to Water Use in the Sediment Sink Zones', *Indian Journal of Landscape Systems and Ecological Studies*, 18(2), 170-207.

— (1996): 'Study on the Natural Hazards of Parts of the Sikkim Himalayan Terrain with Special Reference to Landslides', Proceedings of International Seminar on Disasters and Environment: Monitoring and Forecasting in R.B. Singh (ed.), *Research in Geography*, Delhi University, Vol. II, 285-90.

11

Two Ancient Ports of South India – Muziris and Colchi: A Comparison in Maritime Trade

S. Ebeneezer

Introduction

India has been engaged in maritime activities since the pre-Christian era. The 1st millennium B.C. saw the emergence of many port locations on both the West and East coasts of India. The *Periplus of the Erythraean Sea*, a Greek travel book written in the 1st century A.D. (80 A.D.), lists a series of ports along the Indian coast (Figure 1), such as Barygaza, Suppara, Kalliana, Semylla, Naura, Tyndis, Muziris, Nelcynda, Colchi, Khaberis and Puduca (Schoff, 1912).

Pliny (50-60 A.D.) and Ptolemy (150 A.D.) also make references to many port locations and their maritime activities (Warmington, 1928). Sangam Tamil literature, probably written in the 1st or 2nd century A.D., also speaks about the existence of many ports such as Muciri (Muziris), Tondi, Korkai (Colchi), Marungurpatnam, Saliyur and Puhar.

Figure 1
India: Ports During First Millennium B.C.

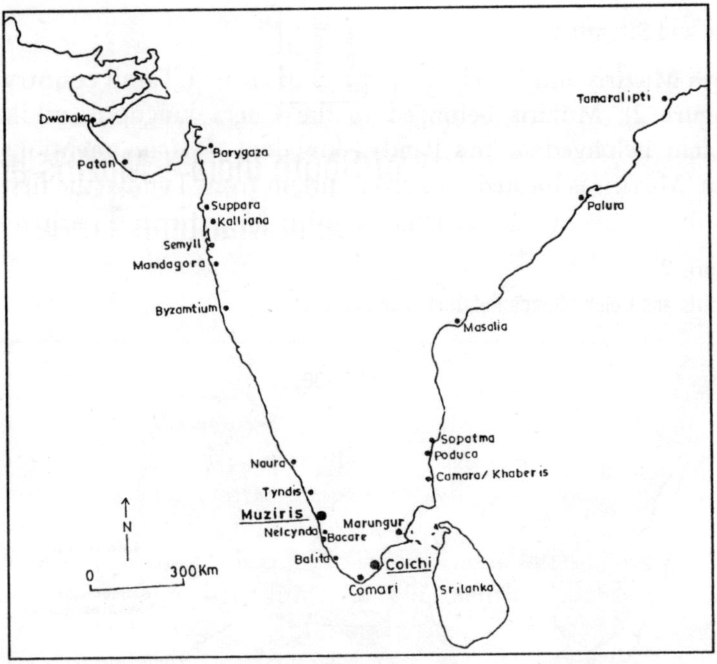

Practically every one of them has declined and, now, it is very difficult even to trace their locations.

While most of these ports were recognised as maritime trade marts by the Greeks and the Romans, only a few of them were considered emporia. Two of them in Southern India have received recognition both in classical European and Indian vernacular literature – Muziris (Muciri/Muyirikodu) at the Periyar mouth on the West coast and Colchi (Korkai) in the Tambraparni delta on the East coast. They belong roughly to the same period and have many comparable features of maritime trade relations and yet differ from each other in some significant respects. The objective of the present paper is to attempt a systematic comparison of the two ports, Muziris and

Colchi, with regard to their site and situation, trade links and the nautical environment in which they evolved.

Site and Situation

Both Muziris and Colchi were parts of ancient Tamil country (Figure 2); Muziris belonged to the Chera kingdom, while Colchi belonged to the Pandya kingdom. *Periplus* mentions that 'Muziris is located on a river, distant from Tyndis (the first

Figure 2

Muziris and Colchi: Space Relations and Linkages

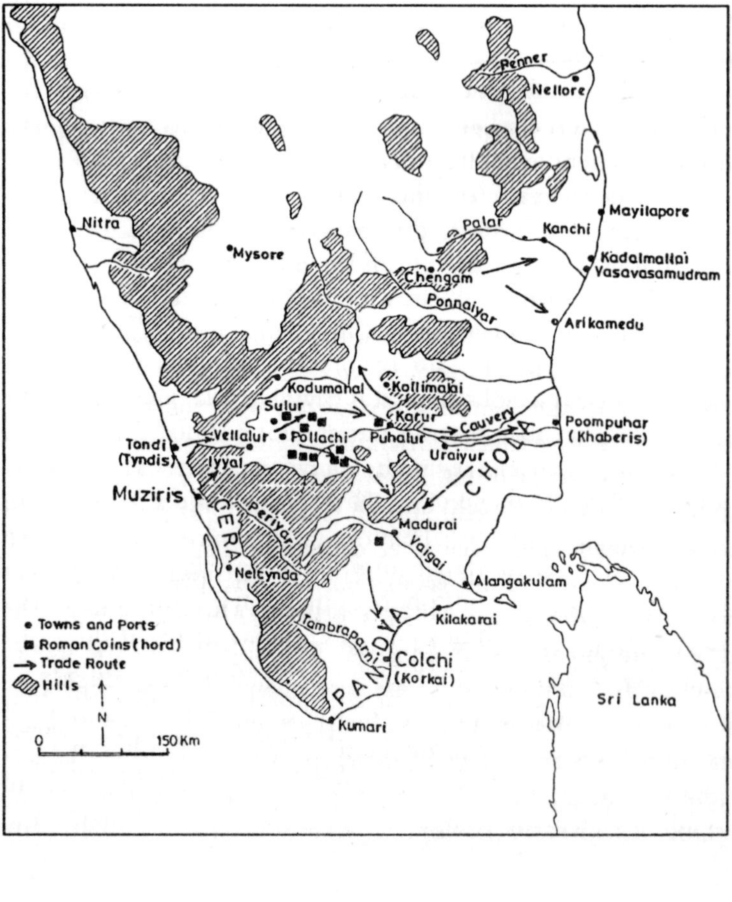

market of Damirica), by river and sea five hundred stadia and up the river from the shore twenty stadia' (Schoff, 1912, 44). Again, the location of Muziris has been linked to yet another port, Nelcynda, which is about 500 stadia away from Muziris by river and sea. So the location of Muziris could by identified only by the relative locations of Tyndis and Nelcynda. While no mention was made in the *Periplus* about the name of the river by the side of which Muziris was located, a Tamil poem of the Sangam Age mentions that Muciri (Muziris) was located at the mouth of "Sulliam Periyaru".[1]

Scholars differ in locating the exact site of these ports. However, by taking into consideration the textual facts and geographical and historico-political situation, many scholars have fixed the location of Muziris at Kodungaloor (Cranganore) at the mouth of the Periyar river (10°14'N., 76° 11'E) (Schoff, 1912). Tyndis and Nelcynda have been fixed at Ponnani and Kottayam, respectively.

Korkai, referred to as Colchi by *Periplus*, Kolkhai by Ptolemy (130 A.D.) and Korkai in Sangam literature, took a long time for its location to be fixed. Many scholars have erroneously identified its location: for example, Wilson Bodan identified it with Kilakarai, a sea port in Ramanathapuram (Sundararaj, 1981). Only in 1838 did an anonymous contributor to the *Madras Literary Journal* bring to light the existence of a place called Korkai, known for its pearl fishery in the Tirunelvelly (Tirunelvelli) district. On this basis, Caldwell conducted archaeological excavations at Korkai and its surrounding areas and identified the village of Korkai with the port city referred to by the Greek geographers and Sangam literature (Caldwell, 1877). Thus, Korkai, the ancient port of the Pandya kings of the Sangam period, a centre for pearl fisheries, has now been identified with the village of Korkai, which lies 6 km away from the coast to the North, off the left bank of the Tambraparni river, in Tuttukudi district of Tamil Nadu.

The circumstances under which the two ports have been abandoned is worth investigating.

The Case of Muziris

The West coast of India had its early trade contacts with Rome definitely from around the time of Asoka, if not earlier. Asoka had himself cited many Mediterranean rulers like Anteachus Theos of Crete, Antigous of Macedonia and Ptolemy Philadelphus of Egypt (Rajan, 1994). In fact, there are references in the *Bible*, that Solomon's ships (in the middle of the 10th century B.C.) carried items like gold, apes, peacocks, spices and ivory. The fact is that these goods were mainly South Indian products and, hence, they might have passed through a port in South India.

In the early days, trade with the Western nations was carried on along the coast of the Arabian Sea, reaching the Mediterranean Sea either through the Red Sea or the Persian Gulf and, thence, by land. The movement by sea was primarily through coastal voyages, hopping from port to port all along, and not directly across the high seas (Thomas, 1931-32). This was a circuitous route, it took a long time and the dangers of travel were great. It was only in the middle of the 1st century A.D. that the navigator, Hippalus, discovered that by taking advantage of the monsoon winds, ships could sail straight across the ocean and reach India quickly (Warmington, 1928). Ships would sail in July from the Red Sea ports and Muziris could be reached in 40 days. The direction of the wind led the ships straight to the Malabar coast and Muziris was the first port touched in India; Pliny calls Muziris the nearest mart in India. Those who wanted to go to other parts of India would then continue their journey by coastal vessels or by land routes. The ships would then take their cargo and sail back to the Red Sea mouth in December when the Northeast monsoon sets in. The discovery of the trade winds gave a fillip to Roman trade and made Rome the leading commercial nation of the world.

Muziris became one of the chief Roman trade centres, evolving into an emporium.

The discovery of the monsoon winds had a great impact on the South Indian political economy, too. The Chera kings initially had their capital near Muziris and took control of the entire maritime activity. Many large Roman ships used to visit Muziris. Pepper, cinnamon, ginger and other spices, which were locally available along the Malabar coast, formed the traditional exports from Muziris, together with beryl obtained from the Kongu country (Coimbatore district), an interior part of the Chera kingdom. The other items exported from Muziris were fine pearls, ivory, quality cotton textiles, silk cloth, ebony, tortoise shell, spikenard – from the Ganges – and malabathrum – from the interior. The imports to Muziris included punched gold coins in large quantities, mainly from the Roman empire, amphorae, ceramics, topaz, crude glass, copper, coral and wine. Some quantity of wheat was also imported for the sailors.

With expanding trade, the competition from the rival Chola and Pandya kingdoms increased. The control over trade and resources must have stimulated the Cheras to have another capital at Vanji (Karur) in the interior part on the banks of the Amaravati, a tributary of the river Cauvery. King Cheran Senguttuvan had complete control of the Kongu country and the second capital, Vanji (Karur), gained more importance than the capital on the West coast (Rajan, 1994). Hence, Ptolemy identifies two Karurs, one on the West coast (Koreoura), and the other in the interior part (Karoura). The capital Karur in the interior was a vantage point, centrally located on the major trade routes (Figure 2).

At least three major trade routes of that period can be recognised:

1. The route that links Mysore (Erumaiyur) with Uraiyur, the Chola capital, via Karur, the Chera capital;
2. The route that links Muziris and Karur, passing through Palghat, Vellalur, Sulur and Kodumanal (the beryl

processing centre), recognised as the well established Kongu Peruvazhi; and

3. The route that links Muziris with Madurai, the Pandya capital, via Pollachi, and running beyond to Korkai.

Roman hoard sites were amply located on the way to Karur as also in and around Karur. Roman artifacts and coins that were unearthed and the spatial location of such sites throw much light on the amount of external trade and maritime contacts. Many of the coins unearthed belong to the period of Augustus (27 B.C. to 14 A.D.) and his successor, Tiberius. A few Republican coins have also been unearthed. It indicates that Indo-Roman trade started in the pre-Christian era around the 3rd century B.C. and reached its zenith during the early Christian era.

Warmington (1928) says that the Roman trade balance during the early Christian era was favourable to India. Sangam literature shows that the Yavana ships came to Muziris with gold and returned with pepper;[2] the gold coins unearthed at various places in the Tamil country strengthen this view.

Muziris continued to be a brisk trading port during the mediaeval period, with a possible shift in site. In later mediaeval days, it was known as Mahodayapuram or Vanji. It was after 1350 A.D. that the port started declining and, when the Portuguese arrived around 1500 A.D., it was struggling for survival. Later, the port became defunct. The archaeological investigation around Kodungallur has yielded cultural material datable to the 9th-11th century A.D. and also carbon dates of artifacts going back to the 4th century B.C. The precise ancient port site of Muziris is still elusive, despite the material and literary evidence. The reasons for its meagre database lies in inadequate understanding of sea-level variations and coastal changes and shifts.

The mediaeval decline of the Muziris complex of ports on the Periyar may be attributed to the sudden deluge in the Periyar in 1341 A.D. and the diversion of its course, which created a drastic change in the marine environment of the port site (Figure 3a). The 1341 A.D. flood catastrophe resulted in the Periyar forming a new left bank channel just upstream of Alwaye and flowing Southwest through Varapuzha. This channel itself further splits into two, of which, a South flowing one enters the Bay of Kochi-Azhi (Kochi) near Tripunithura. The magnitude of this shift in the Periyar discharge can be

Figure 3a
Muziris: Shifting Port Location

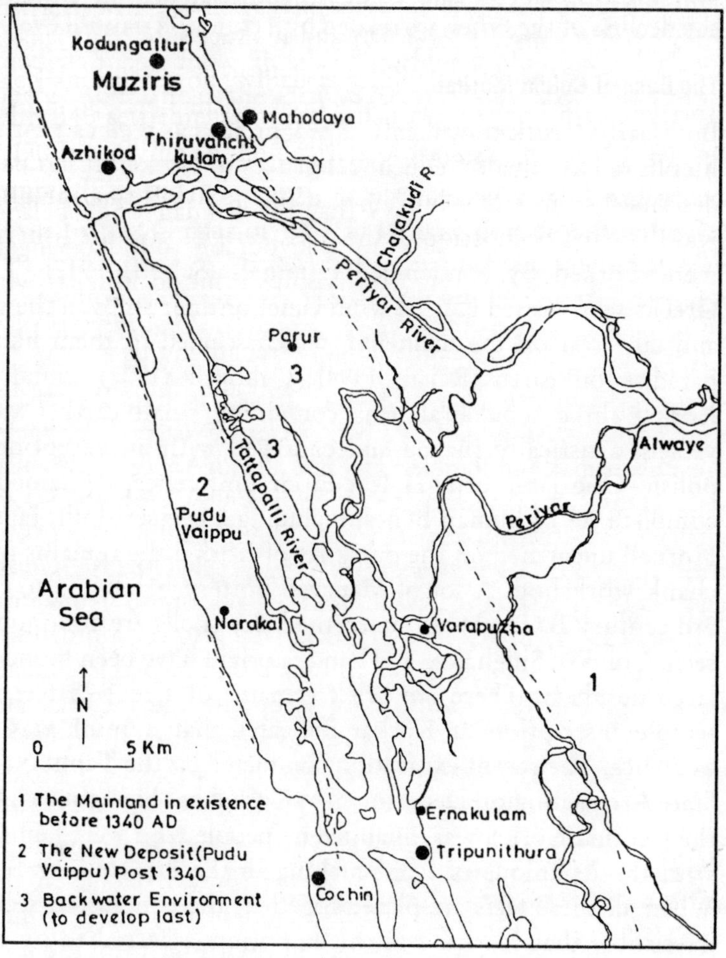

gauged from the fact that, today, of the 12,800 cusecs annual discharge, nearly two-thirds is diverted into the left bank channel, leaving a mere third of the discharge, augmented by the Chalakudi river discharge, entering the sea through the old mouth. This shrinkage in volume itself is a major factor contributing to the decline of the old port, viz. Kodungaloor (Menon, 1902; Sukumar, et al, 1998).

Strangely, the decline of Muziris coincided with the emergence of the new port of Kochi. The rise of one port and the decline of the other are related by a common natural event.

The Case of Colchi (Korkai)

Both early Greek travellers and geographers as well as Sangam literature have made frequent references to Korkai. It has been described as an emporium known for its pearls. Pearl fishing was the chief occupation of the port. *Periplus* says the fisheries were worked by condemned criminals (Schoff, 1912). The Greeks who visited Korkai would anchor their ships in the sea and take rest on the mainland, which seemed to them like a paradise on earth (Rajan, 1994). Caldwell (1877) found at Korkai three sepulchral urns containing small earthenware vessels, artistically shaped and endowed with an exceptional polish. One large urn, 11 feet in circumference, contained a complete set of human bones, including a perfect skull. James Hornell unearthed on the outskirts of Korkai the remains of a chank workshop. A lot of Mauryan pottery belonging to the 3rd century B.C. was also unearthed. Sea shells are commonly seen. Coins of Singhalese and Pandya origin have been found in large numbers. There are brick remains of size 1.25 feet. A temple inscription at Korkai indicates that a mint was in existence. The recent excavation conducted by the Tamil Nadu State Archaeological Department at Korkai has brought to light the fact that Korkai was inhabited by people who were familiar with the technique of iron smelting in the 8th century B.C. When all these facts are placed together, the conclusion seems irresistible that Korkai was the emporium referred to by the

Greeks and one of the great cities of the early Pandyas. As Korkai's activity was mainly associated with pearl fisheries, the major commodity exported was pearls and that imported was horses.

Korkai has been carbon dated to 850 B.C. and it came into prominence in the 3rd century B.C. It lost its importance around the 6th century A.D. as a consequence of the outward progradation of the delta and the shifting course of the river. With the decline of the Korkai port, other port sites such as Kayal (Palaya Kayal) and Punnai Kayal emerged, in turn, as important ports. When Marco Polo visited this part of India in A.D., 1292, Korkai was no more in existence and Kayal port had a prosperous trade with the West as well as the East. Marco Polo says, 'Cail is a great and noble city... It is the city that all the ships touch that come from Hormos and from Kis and from Aden and from all Arabia laden with horses and with other things for sale and this brings a great concourse of people from the country round about and so there is great business done in the city of Cail' (Yule, 305). The Tambraparni delta is characterised by many lagoons and backwaters called 'kayal'. There are many villages having the word kayal as a suffix, e.g. Manjanir Kayal, Palaya Kayal, Punnai Kayal and the like. This has created a difference of opinion among scholars in fixing the Kayal port, referred to by Marco Polo. Some scholars think that the site might be Punnai Kayal, while some others, like Yule (n.a.), fix the location at Kayalpatnam. However, there is no ground evidence in support of the latter. Caldwell (1877), during his excavation at Palaya Kayal, came across numerous pieces of China porcelain and Arabian pottery. He states that Palay Kayal might have been at one time (in the 13th century A.D. at any rate) a prominent port and probably that might be the port referred to by Marco Polo (Caldwell, 1877). The very name 'Palaya Kayal', which means the 'old Kayal', shows that it might have been the original Kayal.

A close examination of the location map of Korkai and its surroundings would bring to light the changing nature of the

Tambraparni river delta (Figure 3b). The port sites around the Tambraparni delta were liable to lateral shifts, either due to natural calamities or due to a changed local sea environment. Thus, Korkai, Kayal, Palaya Kayal and Punnai Kayal, all at the

Figure 3b
Colchi: Shifting Port Location

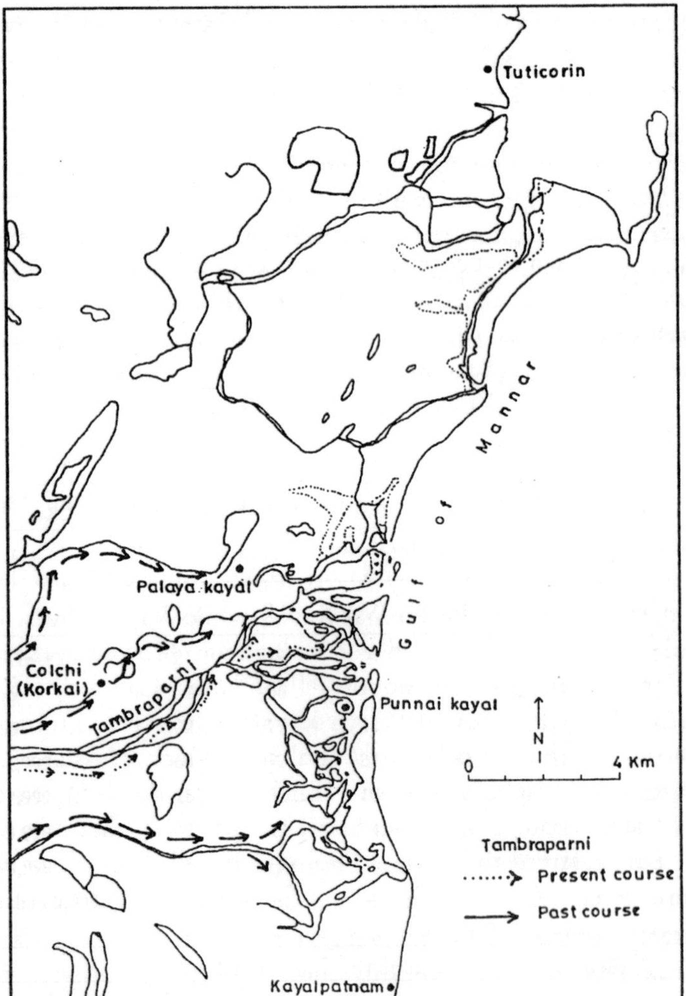

Tambraparni mouth, within a few kilometres of each other, are site adjustments to shifting channels of essentially one port at the mouth of Tambraparni, notorious for flash floods (Arunachalam, 1995).

Palaya Kayal is situated 4 km Northeast of Korkai while Kayalpatnam is 10 km Southeast of Korkai. Punnai Kayal is a little seaside settlement separated by an arm of the Tambraparni. It appears that when Kayal (Palaya Kayal) lost its importance, the Muslim community that lived there shifted to Kayalpatnam while the 'Parathavas' shifted to Punnai Kayal (Fernandes, 1974). Punnai Kayal figured prominently in the affairs of the Portuguese since by that time, Kayal (the present Palaya Kayal) had receded inland. But when these places got silted up in course of time, they were abandoned. Eventually, it led to the development of Tuticorin.

Conclusion

The two ports, Muziris and Colchi, respectively on the West and East coasts of India, are fairly representative of Indian ports in the pre-Christian era. Muziris, like others on the West coast, is a port on a river estuary. The fury of the monsoons made these ports seek a sheltered site, fairly inland, not directly exposed to the open sea and the onslaught of direct waves. Initially, they were on tidal heads, adjoining a safe landing site, where cargo could be loaded and unloaded in all types of weather. With the passage of time, as the estuaries started silting, the port sites moved downstream towards the sea and, ultimately, even became roadsteads. In the case of Muziris, a further accentuating factor was the decline in river discharge after the 14th century A.D.

On the other hand, the East coast ports were on fertile deltas, on prograding coasts. Floods, cyclone vulnerability and lateral shifting in the lower river courses have led to shifts in the port sites, on backwaters or inlets, not downstream, but on sites laterally shifting, seeking flood-free locations. Thus, Korkai is an early location of a port site that has swung laterally along the shores, north or south. Between the two, Korkai appears to be older than Muziris, going to a more distant past as revealed by carbon dating. Both, however, seem to have started declining in the early Christian era, in common with many other sites. As emporia, Korkai's dependence was heavily on its own product, pearls, while Muziris had a wide range of products, though the most significant was *yavana priya* or pepper. Korkai, being on the East coast, appears to have had the benefit of products from further East, Southeast Asia and China.

Acknowledgement

The author wishes to thank Prof. B. Arunachalam for his guidance and encouragement and for sparing his valuable time to go through the manuscript.

Notes

1. Akananuru: verse 149.
2. *Ibid.*

References

Arunachalam, B. (1995): 'Integrated Interdisciplinary Approach to the Study of Ancient Indian Ports: A Case Study of Tamil Nadu Coast', Proceedings of the Seminar on Ancient and Medieval Ports of India, organised by Maritime History Society, Mumbai, 26th October 1995, 18-27.

Caldwell, Rev. R. (1877): 'Explorations at Korkai and Kayal', *Indian Antiquary*, VI, 80-82.

Fernandes, T.I. (1974): 'Kayal Ports of Pandya Nadu', (in Tamil), Araychi, IV(3), 219-24.

Menon, P. (1902): 'Discursive Notes on Malabar and Its Place Names', *Indian Antiquary*, XXXI, 338-50.

Rajan, K. (1994): 'Early Maritime Activities of the Tamils', Proceedings of the International Seminar on Techno-Archeological Perspectives of Seafaring in the Indian Ocean, New Delhi, 97-108.

Schoff, W.H. (1912): *The Periplus of the Erythraen Sea*, Longman Publication, London.

Sukumar, B., Sukumar, A., and Arunachalam, B. (1998): 'Near Shore and Inshore Morphological Changes in the Periyar Mouth, Kerala', Proceedings of the Seminar on Nautical Cartography, organised by Maritime History Society, Mumbai, 28th and 29th March, pp. 80-86.

Sundararaj, T. (1981): 'Rise and Fall of Korkai', *Journal of Tamil Studies*, 55-63.

Thomas, P.J. (1931-32): 'Roman Trade Centres on the Malabar Coast', *The Journal of Madras Geographical Association*, VI (3 & 4), 231-40.

Warmington, E.H. (1928): *The Commerce between the Roman Empire and India*, Cambridge University Press.

Yule, C.B. (n.a.): 'Ser Marco Polo (Translation) II', 305, as in *Gazetteer of Tirunelvelli District*, 433.

12

A Perspective on Solid Waste Management in Bangalore City: Viable Technological Options

Daksha C. Barai

Introduction

Solid wastes refer to discarded materials in a solid form that result from man's activities, which, on their immediate production, are of no use. They may later be brought back into use when recycled or reclaimed. Wastes in urban areas arise from residential, commercial, industrial and institutional uses, which often are not handled properly. Residential wastes are in the form of garbage, including fuel residues, containers and packages, glass bits, food residues and garden wastes. Commercial wastes are generated from shops, offices, restaurants, warehouses and hotels. They range from vegetable wastes in markets to packing material, metal, plastic, textiles or chemicals. Institutional wastes include those wastes that arise from hospitals, educational institutions and from other services ranging from bandages, syringes, bottles, paper and associated materials. Apart from these sources, construction debris with its heaps of sand, stone, bricks, rubble, etc., on streets and vacant plots are major sources of nuisance which also cause traffic hindrances, besides presenting an ugly sight or visual pollution.

The Problem of Wastes in Bangalore

Bangalore city is a rapidly growing metropolis both in terms of population and activities. It has been observed that the city's garbage is doubling every decade. The city generated about 1,012 tonnes of wastes per day in 1984-85; by 1994-95, its wastes had increased to over 2,000 tonnes per day. The city's central location, pleasant climate, parks and playgrounds have attracted a wide range of activities that have added to its garbage. The generation of about 2,300 tonnes of wastes per day is almost over double that in the year, 1984-85. While the population has increased by 130 per cent over the past decade, the waste disposal infrastructure has not increased proportionately; dustbins have increased by 48 per cent, the menial staff by 8.1 per cent, and associated transport by 72 per cent. The shortage of staff and vehicles hampers the efforts of the corporation towards garbage disposal. Hence, garbage is often found spilling over from dustbins to roads and vacant plots, especially after festivals.

Garbage is generally collected by conservancy staff and disposed off to landfills with the help of lorries. Due to the growth of the city, some of the landfills are now within the city limits and some have been closed down. The landfill at Binnypet and on Magadi Road have been filled and closed, while the one at Koramangala has been sealed and converted into an indoor stadium. At present, nine landfill sites have been identified by the government outside the city limits, mostly to the North and South of the city.

Composition of Urban Wastes in Bangalore

The wastes generated in Bangalore primarily consist of fermented matter, along with sand, stone, dust, textile, paper, plastic, etc. (Baud, et al., 1994 a,b). An analysis of these wastes has indicated that municipal wastes averaged 8 per cent paper, 6 per cent plastic, 6 per cent glass and 3 per cent metal. Further, about three-fourths of the garbage consists of organic matter and about two-thirds are recyclable.

As organic matter dominates the composition of solid wastes in Bangalore, there is a greater possibility of its recycling. It can be converted to manure, thereby returning nutrients to nature and enriching the soil. This would be a viable strategy to meet the needs of a garden city like Bangalore, in which manure is needed for the city's parks and gardens. Non-compostable solids will involve an informal system of reuse; household, commercial and institutional wastes pass through a hierarchical network of waste pickers, itinerant buyers, retail buyers, wholesalers, registered buyers and recyclers.

Formal System of Waste Management

The formal system of waste management is handled by the health and engineering departments. While the former lifts the wastes, supplemented by rag pickers who search the bins and other sources for recyclable material, the latter takes care of the maintenance and purchase of vehicles. The first move towards a formal system of waste management involved the sending of wastes to the Karnataka Compost Development Corporation (KCDC) plant at Bommanahalli, the only place where the Bangalore City Corporation (BCC) could legally send its wastes for composting; on an average, 50 tonnes a day was accepted by the KCDC for composting. The compost plant is a joint venture of the Karnataka Agro-Industries Corporation, Karnataka State Co-operative Marketing Federation Limited and the BCC. The original plant, with a daily capacity of 200 metric tonnes, was highly mechanised; however, it went out of order in 1986. The present semi-mechanised plant produces three grades of compost used for agriculture, horticulture, sericulture and forest.

Community-based Formal System

Under the banner of Swabhimana, the core group of the BCC, along with NGOs, deals with a system of garbage management. The first integrated project, Wastewise, involving 500 households in Jayanagar, was initiated by the Mythri Sarna Seva Samithi (MSSS), an NGO that works for street children and in slums. It comprised collection, disposal and recycling – the basic components of waste management. A portion of the public park in the locality was used as the site for composting. A hierarchy of supervisors, coordinators and rag pickers was inducted into the project for collection, disposal and composting. The residents were supplied with bins for wet and dry wastes, and rag pickers were given equipment and trolleys to handle and transport the garbage (Joseph, 1994). The organic matter was converted into compost, which could be sold back to the residents for their kitchen gardens. Wastewise set up similar projects at BTM Layout and Sadashivnagar with about 200 households. Similar projects in solid waste management have been taken up in other residential areas such as Frazer Town, Jayanagar, Rest House, Crescent Park (Farmers' Training Institute along with Agricultural College), Kensington Apartments, Ulsoor, Oakwood Apartments, Koramangala, Indiranagar and Richmond Road. There were also other individual efforts in areas such as Rajajinagar, Indiranagar and Koramangala. The concept of small-scale recycling of wastes has crept into the new large apartment complexes as well. The projects involve the efforts of NGOs such as the Centre for Environmental Education (CEE), the MSSS, BOSCO, REDS, etc., towards biodegradable waste management. Efforts are also being made by institutes such as, the GKVK for residents in the surrounding areas to collect garbage with the help of street children for vermi-composting technology. Training and literacy programmes help in separating and recycling wastes. Efforts are made by hotels such as West End to separate kitchen wastes to set up their own biogas plant to develop power for their kitchens. Localities such as Majestic and the K.R. Market produce large quantities of waste every day and transporting them to dumping grounds is a herculean task. The CEE, an NGO, has launched a scheme to collect the waste from these centres and send them to Coles Park for composting. The GKVK and the University of Agricultural Sciences help in the initial supply of earthworms to the projects and provide expertise and training to households, gardeners and others interested in the technique.

In all these projects, garbage is sorted out at source (the residence). The rag pickers, provided with uniforms, gloves, shoes and socks, help in the collection of garbage and composting. They are paid Rs 500 per month for working six days a week. These wastes are then shifted to line composting pits capable of holding 500 kg of wet waste. Three weeks after processing, a fine brown granular compost is available for sale.

Informal System of Management of Wastes

Bangalore has a production sector from which waste materials are recovered, traded or recycled through a system of informal economy. The producers of recyclable waste, i.e. households, and commercial and institutional establishments, arrange to sell their waste through waste pickers, itinerant buyers, retailers, wholesalers and recycling establishments to produce a hierarchical structure of networks within various groups.

Waste pickers, estimated to range from 20,000 to 30,000 in number, come from the poor, disadvantaged, low-caste sections of the population, who collect about 30 kg of waste to earn barely Rs 20 per day to meet their basic needs.

Itinerant buyers buy or barter waste directly from households, offices or shops to sell them to medium sized waste buyers. Some small buyers sell directly to wholesalers. About

100-200 units supply waste to two or three itinerant buyers, who collect 20-75 kg and earn Rs 50-70 per day.

Waste buyers number about five per neighbourhood and they are functionally close to wholesalers. They pick up about 150 kg to individually earn Rs 25-40 per day. The estimated number of buyers is about 600. These waste buyers purchase bits of plastic, glass, paper and metal. These are sorted out to be sold to specialised wholesalers. Wholesalers sell their materials directly to recycling enterprises. Wholesalers specialising in waste material usually employ casual labourers to sort out and store materials to be sold to recycling units. The majority of the paper wholesalers are located at New Tharagupet, and they provide paper and packaging material to paper factories within the state and in the other states. The glass wholesalers are located at Rajajinagar and in the City Market area and they deal in liquor and medicinal bottles to sell them to the concerned industries. There are about 12 glass wholesalers and three medicinal bottle wholesalers in Bangalore. Plastic is recycled primarily by 15 wholesalers around Jolly Mohalla and K.B. Market. All plastic waste, ranging from plastic carry bags to mugs and buckets, is supplied to these wholesalers.

Recycling Enterprises

Recycling enterprises play an important role in waste management as they help in the local economy and at the same time maintain the environment. They help in waste separation, reducing pollution, saving on raw materials and providing employment to a large number of people.

Bangalore has three recycling units for paper, two for glass and 350-500 for plastic; the paper recycling units use paper, kraft cutting, cotton rags, cigarette stubs and filters, bagasse, packing paper, writing pads and envelopes. Glass recycling units use colourless glass; coloured glass is sent to Mumbai. Plastic can be recycled seven times and a wide range of recycled products, such as pipes, pots and buckets, containers and plastic bags, are available within the city limits. It has been noticed that the lower level functions of waste buyers and itinerant buyers are ubiquitous and distributed throughout the city, but the higher order services of wholesalers and recyclers lie only in the Central and Western parts of the city close to the main market.

Perspectives in Waste Management

The efforts made by NGOs in formal waste management have been explained earlier. These seem to be on the increase as many among the new apartment complexes tend to recycle their organic waste themselves. Today, with the help of the CEE, the Thoreau Foundation, SUCHI, Wastewise and others, the city has been able to recycle almost 300 tonnes of organic waste. The emphasis has so far been on the higher income areas, but there is a need to rope in communities in lower income areas and slums as well. For this, it is essential for NGOs to integrate the schemes through CBO and implement them along with infrastructural back-up by the government.

Some efforts are worth attention in this regard. Excel Industries of Mumbai, in collaboration with the KCDC, has revitalised the plant by using biotechnology to increase compost production. The Engineering Resources Group, in collaboration with the Computer Maintenance Company, has set up a fuel pelleting plant in Mumbai to produce 30 tonnes of pellets a day. A similar project proposal has been submitted to the authorities in Bangalore. The Eighth Five-Year Plan budget has allocated Rs 3 crore to the compost plant in the city. It is essential to draw private organisations into waste management as is done in Ahmedabad. Such sustainable development projects are being considered in the city.

All these schemes of waste management will help keep the environment clean and generate employment to the unorganised sector.

References

Baud, I., Huysman, M., and Schenk, H. (1994a): *New Approaches to Urban Social Waste Management: Linkages between Formal and Informal Systems of Source Separation, Collection and Recycling in Indian Cities*, University of Amsterdam.

Baud, I., and Schenik, H. (1994b): *Solid Waste Management: Modes, Assessments, Appraisals and Linkages in Bangalore*, Manohar Publishers, New Delhi.

Joseph, A. (1994): 'Urban Garbage: Reorganizing the Removals', *The Hindu*, Survey of Environment, Chennai. 183-87.

13

Social Infrastructure: A Study of Housing in Ahmedabad City

K.M. Kulkarni

Introduction

The word 'infrastructure' means many things to many people, yet there is a consensus as regards its laying the basic structural foundations of a society or an enterprise. According to the U.S. Academy of Sciences' National Research Council, infrastructure includes "a wide range of economic and social activities, a crucial enabling environment for economic growth and enhanced quality of life" (McNeil, 1993). There is a strong link between infrastructure and development and progress in infrastructure has to precede, accompany and follow that in development. Infrastructure is classified depending upon the nature of its service, cost, level and extent of operation, users, etc. Some of the major types of infrastructure are: (a) physical and social; (b) public and private; (c) facility and service; and (d) demand point and supply point. This study aims at analysing housing facilities as a part of social infrastructure, which promotes and strengthens social well-being. Though many scholars consider housing a part of physical infrastructure due to its material structure, this study looks at housing from the viewpoint of its types and their socioeconomic and cultural implications.

Housing is one of the basic requirements of civilised society. Initially, a house got developed as a structure for protection from environmental conditions. Although this objective needs to be achieved in the present-day housing facility, there are other requirements, too, such as protection from anti-social elements, greater comfort, privacy, architectural appeal, and occupational and cultural needs. The fulfilment of the housing requirement is an indicator of social development and well-being. At times, lack of adequate housing becomes a constraint in further economic development.

Housing is an expression of culture, the level of technology attained by society, standard of living and development. An average size of a house in terms of built-up area shows a negative relation to size of family as well as that of a city. Shortage of houses increases with city size as does the diversity in terms of house types. Owning a house, especially in urban India, is one of the major aims of the urban middle class. Though the urban areas show a dominance of industrial, commercial, administrative and service activities, the residential function is by far the highest claimant of urban space. This is mainly because of the changing family structure in such areas, from large, joint families to small, nuclear families, where individual family members experience more freedom, comfort and happiness. This needs more housing stock and the rate of housing construction does not keep pace with the demand. Thus, the gap between demand and supply widens with the passage of time. The growth of the city adds to this pressure.

Objectives and Methodology

The study area is confined to the administrative boundary of Ahmedabad Municipal Corporation (AMC) with 43 wards, covering an area of 190.89 km^2 and a population of 28,76,710

(1991). This study deals with an in-depth analysis of housing as a social infrastructure facility in Ahmedabad with a focus on intra-urban variations. The objectives of the study are:

1. To analyse the intra-urban disparities in housing;
2. To regionalise city space based on major house types; and
3. To estimate the housing deficit by wards.

The data on housing were obtained from the town development department of the AMC. The Ahmedabad-Gandhinagar City Map of Setu Publications was used to locate areas with different house types. The data have been processed and tabulated. Simple cartographic techniques have also been used to represent them visually.

Housing Stock in 1996

Eastern Ahmedabad has the highest concentration of housing stock (Table 1) and it is particularly so in the Eastern

Table 1
Distribution of Total Housing Stock, 1996

Micro Urban Zone	*No. of Slums*	*No. of Chawls*	*Slum Houses*	*Chawl Houses*	*Other Houses*	*Total Housing Stock*
Central	65 (6.4)	47 (3.3)	5152 (5.6)	1780 (1.3)	58,010	64,942 (10.6)
Northeastern	149 (14.6)	299 (21.2)	10,551 (11.5)	33,536 (24.5)	48,262	92,349 (15.0)
Eastern	249 (24.3)	850 (60.3)	18,753 (20.4)	87,428 (63.9)	1,29,273	2,35,454 (38.3)
Southeastern	333 (32.5)	105 (7.5)	32,031 (34.8)	7,018 (5.1)	66,713	1,05,762 (17.2)
Northwestern	127 (12.4)	97 (6.9)	15,239 (16.5)	6,476 (4.7	41,283	62,998 (10.2)
Western	28 (2.7)	2 (0.1)	2,669 (2.9)	337 (0.3)	22,998	26,004 (4.2)
Southwestern	72 (7.0)	9 (0.6)	7,726 (8.4)	198 (0.1)	19,880	27,804 (4.5)
Total	1,023 (100.0)	1,409 (100.0)	92,121 (100.0)	1,36,773 (100.0)	3,86,419 (100.0)	6,15,313 (100.0)

Source: Town Development Department, AMC, 1995-96.
Figures in parenthesis shows percentage.

micro-zone. The latter, with its concentration of industries, also has the maximum proportion of *chawls* and *chawl* houses (Bayanwala, 1999). The Southeastern zone, comprising the Behrampura and Isanpur areas, has a third of the total number of slums of the city and a slightly higher proportion of slum houses (Shah, 1978). Housing stock is relatively low in the Western and Southwestern micro-zones. The Central zone, the oldest settled area, is saturated and, therefore, has limited space.

Table 2

Zonewise Percentage of Slum and Chawl Houses to Total Housing Stock, 1998

Micro Urban Zone	*No. of Dwelling Units per Slum*	*No. of Dwelling Units per Chawl*	*Slum Houses (%)*	*Chawl Houses (%)*
Central	79	38	7.93	2.74
Northeastern	71	112	11.43	36.31
Eastern	75	103	7.96	37.13
Southeastern	96	67	30.29	6.64
Northwestern	120	67	24.19	10.28
Western	95	169	10.26	1.30
Southwestern	107	22	27.79	0.71

Source: Town Development Department, AMC, 1995-96.

Slum houses show a predominance in the Southern zones, while *chawl* dwellings are more prominent in the Eastern zones (Table 2). The *chawl* type of housing shows a close association with the textile industry. During the evolution of this industry, *chawls* were constructed to provide accommodation to a large number of workers within a walking distance. Therefore, *chawl* housing is common in the Eastern and Northeastern micro urban zones. The average size of the *chawls*, in terms of number of dwelling units, is also large in these areas. There are only two *chawls* in the Western zone, with a total of 337 dwelling units, taking the average to the highest. Similarly, the average number of houses per slum is higher in the Northwestern and Southwestern zones. All the other zones, except the Central and Western, have a relatively high percentage of poor housing stock, i.e. slum and *chawl* houses and this is particularly so in Eastern Ahmedabad, reflecting on the low socioeconomic profile of the residents there. There is a considerable variation in the percentage of *chawl* houses, ranging from 0.14 per cent in the Southwestern zone to 63.92 per cent in the Eastern zone (Table 1). Relatively less variation is observed in the case of slum houses, 2.90 per cent in the Western zone to 34.77 per cent in the Southeastern zone. This shows that the proliferation of slums is practically ubiquitous, depending on the availability of undeveloped land.

Old Houses

The old houses include three broad categories: traditional well-built houses, *chawl* houses and slum houses. They have been grouped into nine classes on the basis of varying combinations of the three types. The areal groupings have been identified with the help of a triangular graph (Robson, 1979) (Figure 1). Their distribution has been mapped (Figure 2).

Group 1 is characterised by a predominance of (90-100 per cent) traditional well-built houses. These are found primarily in the Walled City, which is the old settled area. There are extensions on the other side of the Sabarmati river opposite the Walled City. Their presence in the peripheral zones in the East and Southeast indicates the amalgamation of old village cores with the expanding city.

Group 2 also shows the prominence of traditional housing but other types of dwellings, particularly slum dwellings, also make their presence felt. Such a grouping predominates in the peripheral parts of the Walled City as well as extensions to the West and South. *Chawl* and slum dwellings are relatively new additions here.

Group 3 shows a slight decline in the predominance of traditional houses (65-80%). Among others, *chawl* houses are

Figure 1
Ahmedabad City: Grouping of Wards Based on Old Dwelling Units

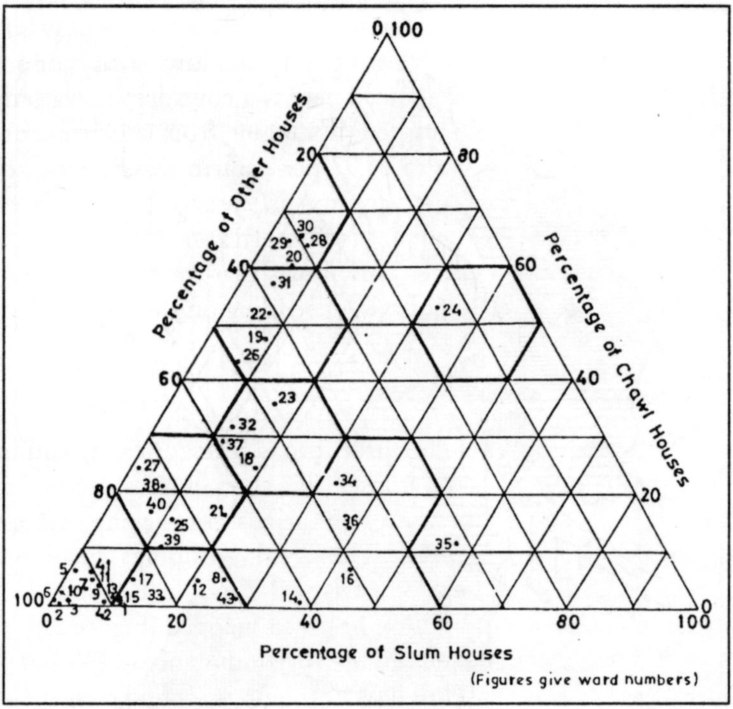

slightly higher in proportion as compared to slum houses. The Northeastern and Northern peripheries, close to the industrial areas, are typical representatives of this group.

Group 4 shows further reduction in the traditional housing component (48-62%); this is compensated by increase in *chawl* houses and slum houses. It is a characteristic of the Northern periphery of the city on both sides of the river. The Southeastern part of the city also shows this combination. The areas are less preferred for residential purposes because of the thermal power station and other industrial activities there.

Group 5 is marked by a high proportion of *chawl* houses (45-60%) along with moderate traditional houses; slum

Figure 2
Ahmedabad City: Distribution of Old Dwelling Units

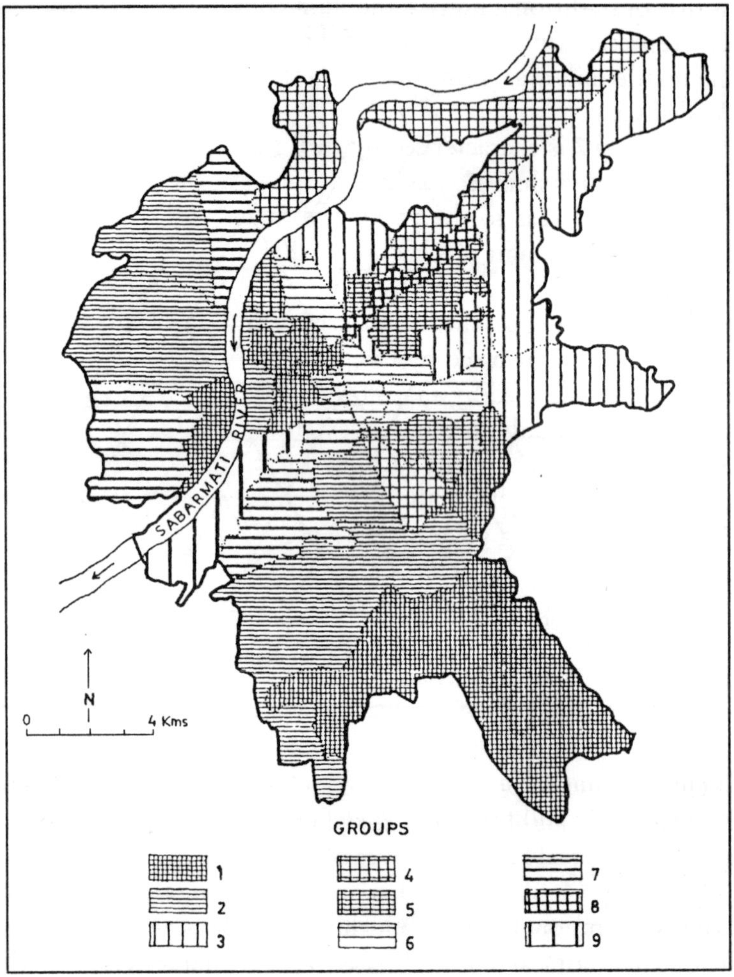

dwellings are limited in proportion. This is a characteristic of the Northern, Northeastern and Eastern parts of the inner periphery. Access to rail appears to be the major consideration.

Group 6 shows a further increase in the proportion of *chawl* housing (60-70%), along with a corresponding reduction in that of traditional houses; the slum component remains low. The representative areas of this group are found in the heart of the textile mill area, immediately to the East and North of the Walled City.

Group 7 is characterised by an almost equal proportion of traditional and slum houses; the proportion of *chawl* houses is limited. The areas to the South and Southwest of the Walled City predominate in this grouping; due to their poor environmental quality, these areas are not preferred for residential purposes and, hence, the higher proportion of slum houses.

Group 8 is marked by the predominance of *chawl* houses (over 50%) followed by that of slum houses (30%); traditional housing is seen only to a limited extent. Spatially, this grouping is restricted to the neighbourhood of rail tracks and the poor quality housing reflects the low socioeconomic status of the residents occupying government land.

Group 9 is characterised by the predominance of slum housing with a moderate proportion of traditional housing. *Chawls* are very poorly represented in the group. This combination is typical of Behrampura to the South, which is a low lying area along the Sabarmati river, with a sewerage farm located therein; the area is thus the least preferred on environmental grounds. The traditional houses had obviously come up before the location of the sewerage farm.

Modern Houses

The Central zone (the Walled City) has old, traditional, huddled, compact houses in the form of *pols, sheris, khanchas, wadis*, etc. About 11 per cent of the total housing stock of the city is found in the Central zone (Tables 1 & 2). Spacious house such as row houses, bungalows and apartments either do not exist or are rarely found in the Central zone. Bungalows and row houses as house types indicate a relatively high socioeconomic status and these house types are found mainly in the Western parts, more particularly in the Western micro-zone and the Northern part of the Western zone (Table 3). The Western zone also has a predominance of apartment structures; the Southwestern and Northwestern zones are next in importance in this respect. The cooperative movement in the urban housing sector is a post-independence phenomenon and the city has experienced a proliferation of cooperative housing societies with structures in the form of multi-storeyed buildings and duplex houses. The Western parts of the city are the most favourable location for such societies. The house type category, 'Others' includes a wide range, such as parks, quarters, towers, enclaves, avenues, palaces, plazas, estates, terraces, complexes etc. The house type of this category, which reflects a higher level of living and taste, dominate in the Western parts.

Table 3
Modern House Types by Micro Urban Zones, 1998

Micro-zone	*Apartments*	*Societies*	*Row Houses*	*Bungalows*	*Others*	*Total*
Central	6 (0.17)	46 (2.22)	-	1 (0.65)	28 (3.38)	81 (1.21)
Northeastern	144 (4.04)	139 (6.70)	-	4 (2.61)	61 (7.36)	348 (5.22)
Eastern	153 (4.29)	269 (12.96)	-	2 (1.30)	77 (9.29)	501 (7.51)
Southeastern	444 (12.46)	152 (7.33)	-	-	82 (9.89)	678 (10.16)
Northwestern	716 (20.09)	686 (33.10)	19 (38.78)	57 (37.25)	204 (24.61)	1682 (25.22)
Western	1230 (34.51)	438 (21.11)	28 (57.14)	85 (55.55)	277 (33.41)	2058 (30.85)
Southwestern	871 (24.44)	345 (16.63)	2 (4.10)	4 (2.61)	100 (12.10)	1322 (19.82)
Total	3,564 (100)	2,075 (100)	49 (100)	153 (100)	829 (100)	6,670 (100)

Source: Ahmedabad and Gandhinagar City Map, Setu Publications, Ahmedabad, 1999. The figures in parenthesis shows percentage.

Living in apartments has become the order for middle class urban residents in Ahmedabad; about 53 per cent of the modern housing is in this form (Table 4). About 31 per cent is in the form of low-height buildings, forming societies or *nagars*, and another 12 per cent constitute 'other' houses. Bungalows and row houses form a very small proportion. While the predominance of apartments outside the Central zone is understandable, that of 'others' in the same zone is due to the presence of varied social groups in the area; their occurrence is also related to urban renewal. In fact, the house types in this zone show a considerable diversity that includes *mohallas*, *manzils*, mansions, *havelis*, *villas*, *wadis*, etc.

Table 4
Intra-Zonal Variation in House Types, 1998 (%)

Micro-zone	*Apartments*	*Societies*	*Row Houses*	*Bungalows*	*Others*	*Total*
Central	7.41	56.79	-	1.23	34.57	100
Northeastern	41.38	39.94	-	1.15	17.53	100
Eastern	30.54	53.69	-	0.4	15.37	100
Southeastern	65.49	22.42	-	-	12.1	100
Northwestern	42.57	40.78	1.13	3.39	12.13	100
Western	59.77	21.28	1.36	4.13	13.46	100
Southwestern	65.89	26.1	0.15	0.3	7.56	100
Total.	53.43	31.11	0.73	2.29	12.43	100

Deficit in Dwelling Units

The deficit in dwelling units for micro urban zones is computed by taking the difference between the number of households and that of houses. Though the data pertain to the 1991 Census, the intra-urban variations are significant. The overall absolute deficit in houses in the city is 3,473 (Table 5). The intra-urban variation is from 62 in the Western zone, where the rich live, to 2,189 in the Eastern zone, which provides refuge to the poor. This means that the shortage of houses affects the low income group primarily. The Walled city also shows a deficit in houses and this is primarily due to the scarcity of housing space and the resulting high rents, making a separate house unaffordable, especially for the poor. Housing deficit is thus an indicator of either houseless population or the sharing of existing houses by nuclear households, relations, paying guests or even sub-tenants, leading to internal crowding (Kulkarni, 1984). Moreover, the slum dwellings and even some of the *chawl* dwellings do not strictly conform to standard housing norms, aggravating the problem further.

Table 5
Intra-Urban Deficit of Housing Units

Micro-zone	*No. of Households*	*No. of Houses*	*Absolute Deficit*	*Percentage*
Central	69,489	68,839	650	0.94
Northeastern	79,826	79,681	145	0.18
Eastern	2,00,473	1,98,284	2,189	1.09
Southeastern	86,013	85,738	275	0.32
Northwestern	65,831	65,745	86	0.13
Western	19,867	19,805	62	0.31
Southwestern	30,665	30,599	66	0.22
Total	5,52,164	5,48,691	3,473	0.63

Concluding Remarks

Housing is an expression of culture and level of technology. In the urban context, it also reflects on the living environment, leading to segregation of society on the basis of socioeconomic status. The history of settlement also plays an important role, giving rise to some differentiation. All these lead to spatial variations in both the stock and quality. The eastern part of Ahmedabad has the highest concentration of housing stock, with a peak in the Eastern micro-zone. However, the same area is characterised by poor housing, with over 90 per cent of the city's *chawl* dwellings and over two-thirds of its slum dwellings.

It is also an area of high absolute housing deficit. Western Ahmedabad, on the other hand, has relatively low housing stock, but many of the houses are modern and, thus, qualitatively superior. The housing deficit is also limited in both absolute and relative terms. The Central zone, occupying the Walled City, falls in between the two, being neither too high in housing stock nor in quality, but not too low either. Moreover, the housing deficit here is limited, despite high land values. Historically, outside the Walled City, the settlements first emerged in the Eastern part on consideration of accessibility, but the settlers were middle class residents. The Western parts are relatively recently settled, after improvement in their accessibility. Being environmentally more attractive, they have lured upper class residents. Although slums have emerged everywhere, they are predominant in the environmentally average Eastern parts or poor Southern parts on the downstream side of the river Sabarmati.

One does not find a glaring deficiency in housing stock as of today. However, it is important to remember that the provision of housing infrastructure is not a static phenomenon. A broad vision, therefore, becomes necessary while planning future housing stock, keeping in mind the growing needs, changing tastes and problems related to other infrastructure in the next 50 years. Such considerations alone would reserve enough space for growth and expansion. Only then can the city develop in a sustainable manner.

References

Bayanwala, V. (1999): 'The Growth of Urban Infrastructure in Ahmedabad and Its Influence on Level of Living', *Nagarlok*, XXXI (3), 34-48.

Kulkarni, K.M. (1984): *Geography of Crowding and Human Response – A Study of Ahmedabad City*, Concept Publishing Company, New Delhi.

McNeil, M. (1993): 'The Changing Nature of Infrastructure', *The Urban Age*, 1(3), 1-6.

Robson, B.T. (1979): 'Housing, Empiricism and the State' in D.T. Herbert and D.M. Smith (eds), *Social Problems and the City – Geographical Perspectives*, Oxford University Press, Oxford, 66-83.

Shah, K. (1978): 'Housing for the Urban Poor in Ahmedabad' in A. de'Souza (ed.), *The Indian City*, Manohar Publications, New Delhi, 145.

14

Perspectives on the Vegetal Cover of North Konkan

Sudha Srivastava

Introduction

The prosperity of a nation lies in the wise use of its natural resources. It is essential that the resources of the biosphere be used in a rational manner and that their conservation be practised on the basis of scientific knowledge, as they are intimately linked with social and economic development. The human relationship with the environment should be based on an approach to development that does not compromise resource renewal. National development strategies and programmes should include balanced rural development for which utilisation of resources should be done in an optimal manner. Over the last couple of decades, scientists have learnt to move the units of heredity, genes, creating new life forms. The biotechnology revolution is set to dominate and alter the world economic scene. Standing at this threshold, an inventory and repository of our biological resources and reserves must be complete and well understood.

The vegetation of a region forms an important renewable resource base. In the tropics, its fortitude lies in its biodiversity, heterogeneity and gregariousness. Since forests are ecological entities, they provide a number of environmental services which include carbon sequestration, maintenance of biodiversity and global and local climatic balance. They are responsible for the retention of the water regimes of the topsoil and its productivity. But these are being skilfully manoeuvred and manipulated for commercial gains, changing the character of the forest altogether.

The present natural vegetal cover of the Indian peninsula is not the original primary climax forest cover. Being inhabited by humans over millennia, the forest cover in several areas has been converted to variations of the original composition or replaced by secondary forests, manmade grasslands or scrubs.

This study focuses on the vegetal cover of the Northern part of Konkan, the coastal belt of Maharashtra lying between the rivers Damanganga in the North and the Savitri in the South (Figure 1). Administratively, it falls in the two districts of Thane and Raigad. Although, physically, Greater Mumbai is a part of this region, it is kept out of the study on account of its highly urbanised and metropolitan character with very little surviving natural vegetation. This paper initially examines the spatial distribution of the natural vegetation cover of North Konkan and its relationship to the prevailing relief. The present status of the vegetation is then investigated and measures suggested for its improvement.

Data Base and Methodology

Survey of India toposheets at scales of 1:25,000 and 1:50,000 provide excellent baseline information. Black and white and false colour composite dia-positives were used to ascertain changes as well as to obtain the most recent picture. The Working Plans and data compiled by the Forest Department helped in understanding the nuances of the spatial vegetation

Figure 1
The Study Region

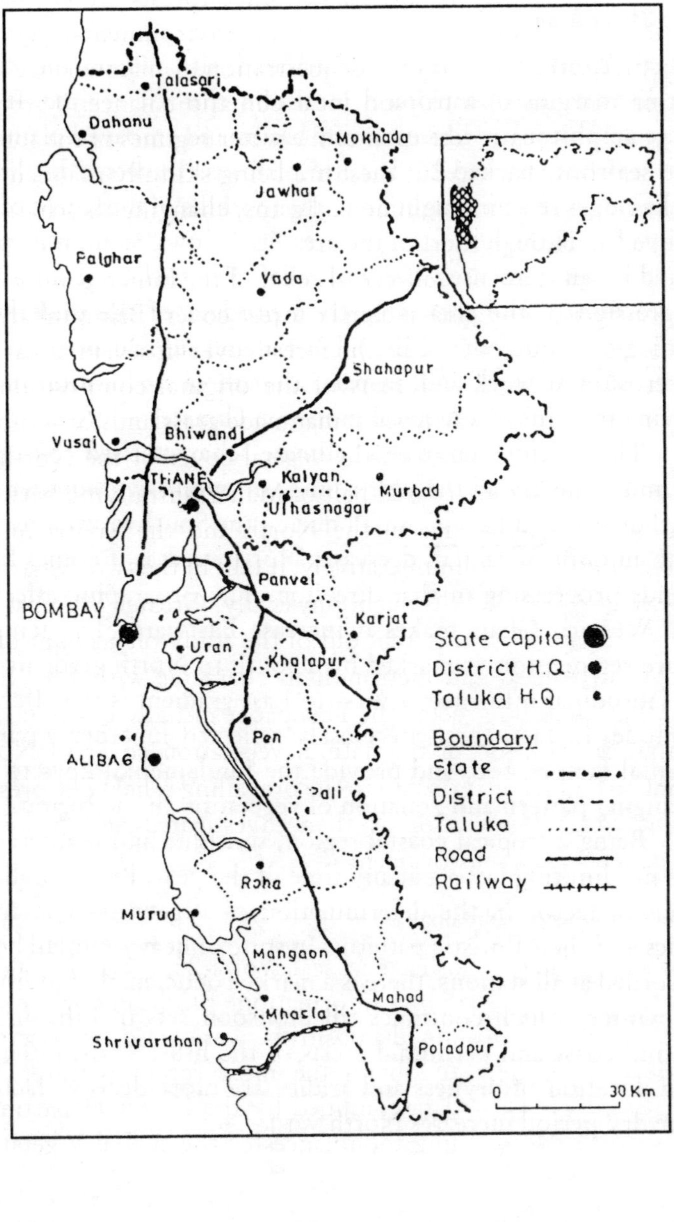

pattern. Field studies were carried out to check the ground truth.

The Study Area

North Konkan is a non-laterised area, situated on the desiccating margins of a tropical monsoon climatic regime. It has large amphitheatre-like estuarine basins and embayed plains on the seafront, backed by basaltic plateaus, traversed by linear dyke ridges reaching right up to the foothills of the scarps of the Sahyadri. Though most of the area lies below 250 m, it is variegated by an immense variety of relief. The surface geomorphic expression of the area is partly a product of the underlying geology, an outcome of the sub-aerial and submarine processes of erosion at work and, in part, a result of its location in the drying margins of a tropical monsoon climate.

The region enjoys a climate characterised by high humidity nearly all the year round, an oppressive summer and well-distributed heavy rainfall during the Southwest monsoon. The amount of rainfall decreases Northward as it comes from winds progressing in that direction. The orographic effect of the Western Ghats makes it increase Eastward. The temperature regime, too, is marked by a South to North gradient due to latitudinal influence; a West to East gradient is the effect of altitude. In fact, these are closely reflected in other environmental factors, too, and provide the fundamental keys to the changing pattern and zonation of vegetation in the region.

Being a tropical coastal region, sunshine and temperature are not limiting factors at any time of the year. But rainfall is a decisive factor in the determination of vegetation characteristics and their floristic patterns. In spite of heavy rainfall being recorded at all stations, there is a marked deficiency of moisture in winter, which continues till monsoon sets in fully in the humid coast and perhumid crests of the hills. Thus, intensity and duration of dryness and aridity are more decisive factors. The dry period increases Northward.

The Natural Vegetal Cover

Although the expression of the natural vegetation is primarily governed by climatic conditions at the micro-level, terrain conditions like slope and elevation play an important role in determining the distribution of particular species and genera; the composition is also influenced by anthropogenic factors. The vegetation complex of North Konkan does not conform to a climatic mono-climax, but is an array of poly-climaxes, an outcome of various physical and cultural forces at work. Konkan forms a distinct botanical sub-unit of Malabar, lying in continuation with it (Meher-Homji, 1987). Forests are generally relegated to higher grounds and steep slopes as cultivation follows the lines of drainage network. Clearing of land for cultivation and other uses has led to the degradation of forests, resulting in dense forests getting converted to an open variety. Further degradation leads to a deterioration in quality of tree growth and a reduction in canopy density, the trees being replaced by scrubs.

Coastal vegetation is characterised by mangroves growing in saline mud flats or in semi-muddy estuarine areas, where the land gently grades into the sea (Figure 2). They are lush green throughout the year, although they may grow in situations that are physiologically dry. Strand vegetation occurs shoreward of the coastline and along the banks of rivers joining the creeks. These continuous belts of thick strands of trees lie just behind the mangroves on the landward side in a number of places. A definite zonation of natural vegetation is seen along coasts, manmade plantations serving occasionally as wind brakes, shielding the coastline as a formidable wall. Casuarina and, of late, eucalyptus are used for this purpose.

Pastures and grazing lands occupy a proportionately small area and are more prevalent in Thane than in Raigad. They represent a stage of transition between marine and terrestrial regimes, sometimes deliberately maintained. It may be noted that grasslands in small, scattered patches or as undergrowth in

Figure 2
North Konkan: Land Use and Land Cover

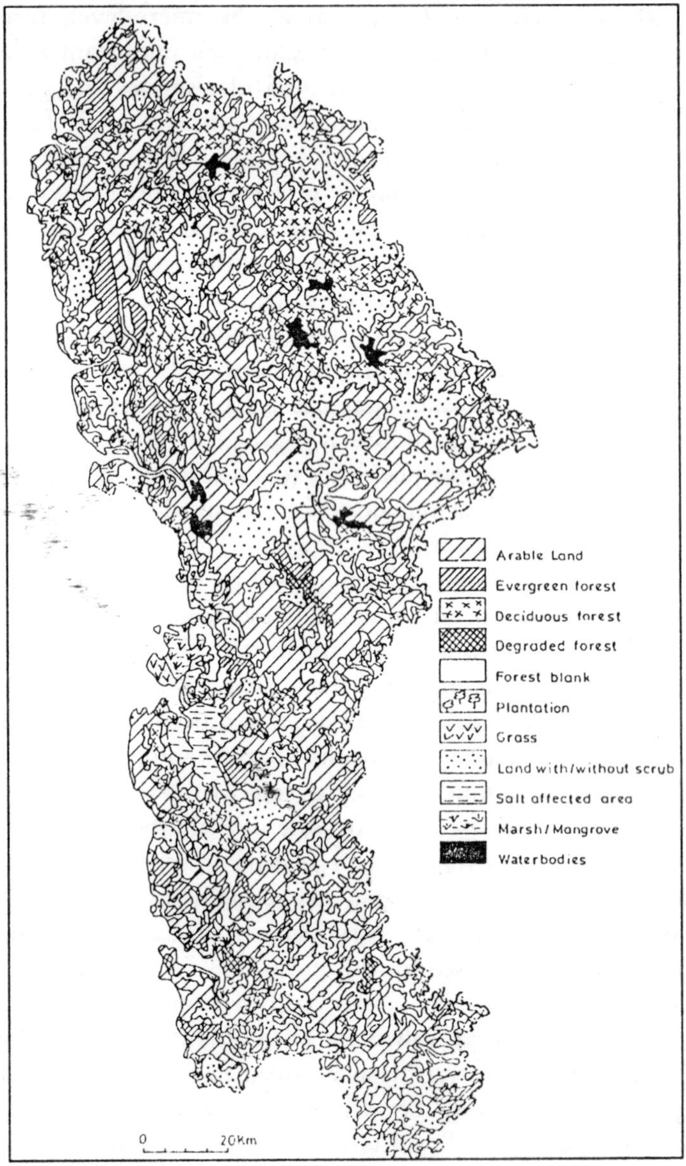

forests are not conspicuous in small-scale imageries and, hence, do not appear prominently on the map. Clearance of forests has increased catastrophically in the last 100 years or so, exacerbating aridity, affecting physiology and the genetics of graminaceous plants. The annual rhythm of water stress, as determined by seasonal patterns of water balance, results in the transformation of forests into grasslands. High rainfall grasslands are found on the Matheran and Dhak plateaus, usually above 600 m where rainfall exceeds 550 cm. Thane has been an important source of fodder to Mumbai. Every railhead on the Western Railway between Virar and Surat has a hay godown.

Forests of the area may be classified as evergreen/semi-evergreen and deciduous. The prevailing climatic conditions result in the development of vegetation through various seral stages into ultimate or climax vegetation, consisting of evergreen forest cover. However, due to a long history of use and abuse/misuse, evergreen and semi-evergreen forests occur only in patches. The natural vegetation of the area lies mainly on red or black soils and laterites. It is primarily moist deciduous forest in which the trees are tall and have a large girth. Density of vegetation varies between 40 and 90 per cent. Height of the dominant stratum in the North is 15-25 m, in the South, it exceeds 25 m with three levels of canopy and an abundance of species. Basically, two vegetation associations belonging to South Indian Tropical Moist Deciduous Forest of Champion's 1936 classification can be identified as follows:

(1) *Tectona* (Teak, Sag) – *Lagerstroemia* (Nana) – *Dillenia* (Karmal) – *Terminalia* (Ain) association on the Western side of the Ghats towards the South, where precipitation varies between 2,600 and 3,600 mm and a dry season of seven months prevails.

(2) *Tectona* (Teak, Sag) – *Terminalia* (Ain) – *Adina* (Hedi) – *Anogeissus* (Dhavda) association towards the North, which is a transition between the moist deciduous and dry deciduous varieties. Here, precipitation is between 1,800 and 2,500 mm and a dry season of eight months prevails.

The Ulhas valley forms an ecotone, North of it being better wooded, while the area South of it has more of scrub vegetation. Shifting cultivation dominates South of the Savitri. Matheran has *Memecylon* (Anjan) - *Syzigium* (Jamun) - *Actinodaphne* (Pisa) association.

True evergreen forest runs in a strip at an elevation of 450-1,370 m to the South of Mumbai. This dense forest has a lofty upper storey, which reaches 35-45 m, occasionally even 60 m. It provides deep shade and conditions conducive to the growth of epiphytes. Buttressed tree trunks are common. Diversity is expressed by a large number of species. Riparian evergreens are found at low altitudes, too.

The semi-green variety is found along the lower slopes of the Western Ghats. Though not as tall as true evergreens, they are nevertheless rich in lianas and epiphytes. About 30 per cent of the trees of the upper stratum are deciduous.

The deciduous or monsoon forests are characterised by leaf shedding for six to eight weeks with a great inter- and intra-specific variation in the leaf fall period. The forest is never completely leafless.

Scrubs reflect a retrogressive stage in the degradation of natural forest vegetation. Thus, they are found within forest tracts, near cultivated lands and in the transitional zones between forests and fields. They may be considered ecological indicators as they result due to the interventionist attitude of humans and their activities in forest lands. These are potential areas for reforestation/afforestation schemes. However, they may succumb to the increasing pressure for diversion/ conversion to other land uses.

Natural Vegetation in Relation to Elevation and Slope

To get a deeper insight into the prevailing relationship between land cover/land use, especially forests, and aspects of relief, profiles were drawn across the entire stretch of North Konkan at intervals of 5' latitude with topo-maps as the base (Srivastava, 1995). From an analysis of these profiles and maps, the following inferences may be drawn. The main categories of land cover/land use are cultivation, forests, scrubs and grassland in the percentage proportion of 43.8:34:19.6:0.8 with a marginal proportion under mangroves and plantations. There is a clear relation between elevation levels and land cover/land use (Table 1, a-c). Cultivation necessarily dominates flat lands or gentle slopes at varying elevations. However, there is a clear preference for heights below 25 m. Above 100 m, there is a sharp decrease in cultivation, while beyond 550 m, it is not practised at all.

Hills and hillocks, flanks of mesas and buttes are all under some kind of forest cover, especially along their middle and upper slopes. However, forests may be found even lower and on gentler gradients. Some amount of forest occurs in all height classes. Forests flourish only when they are under the control of the Forest Department. About 78 per cent of forests are dense jungles, the rest being open. Once a main teak resource region, now only 19 per cent of these forests show a dominance of teak. About 26.4 per cent of dense forests are found between 150 and 250 m, and 21 per cent between 50 and 100 m. Nearly 30 per cent of open forests are found between 150 and 250 m, and 26.4 per cent between 50 and 100 m. Beyond 550 m, open forests are not observed. Teak is an important component in the association of species of both dense and open forests. Its importance stands out (a) between 19°10'N and 19°55'N where teak shows universal presence, especially in the coastal hills, the hills of the central region and plateau areas, and (b) between 18°25'N and 18°40'N, where dense teak forests in the central hills extend right up to the foothills of the Sahyadri. Occasionally, teak grows in association with bamboo. Teak is more significant in dense stands between 50 and 250 m. It must be mentioned that above 550m, teak is totally absent.

Scrubs are scattered widely as they rarely occur as contiguous stretches, but are interspersed with other types of land cover/land use. They are maintained for fodder, firewood or '*rab*' operations. Scrubs are dominant at relatively lower elevations adjoining cultivated lands, 23.5 per cent being found between 50 and 100 m. They tend to occur at all levels. Accessibility to forest cover is the major factor for their existence. It is a secondary sere beginning after succession has been arrested due to disturbance.

Grass, cultivated or growing naturally, either independently or in association with scrub, is more important North of 19°30'N, in coastal areas below 25 m, and also between 50 and 100 m. Some amount of grass cover is found at higher elevations, too.

Table 1

(a) Relation between Vegetation and Elevation Levels

Vegetation Type→ Height (in m) ↓	*Cultivation*	*Scrub*	*OJ*	*DJ*	*OJ (Mainly Teak)*	*DJ (Mainly Teak)*	*Grass*	*S+G*	*Mg+ Ma+ P*	*Total*
<25	218.7	39.3	0.9	7.9	-	1	3.4	1	25	297.3
25-50	110.9	43.6	5.9	24.2	-	204	1.8	1.2	0.6	190.6
50-100	166.5	71.9	27.9	66.8	3.4	20.3	4.3	1.5	0.9	363.2
100-150	47.3	43.9	22.5	55.5	1.7	20.2	2.6	-	-	193.7
150-250	60.6	55.8	34.8	78	0.4	31	0.7	-	-	261.2
250-350	41.9	24.1	12.2	43.7	-	12.8	-	-	-	134.2
350-550	37.6	22.1	7.8	33.7	-	7.6	-	-	-	108.6
550-750	-	4.8	-	7	-	-	-	-	-	11.7
>750	-	-	-	1.3	-	-	-	-	-	1.3
Total	683.5	305.5	112	318.1	5.5	95.3	12.8	3.6	26.4	1561.8

Note: Values are in km.

(b) Percentage Distribution of Vegetation Types in Each Elevation Band

Vegetation Type→ *Height (in m)* ↓	*Cultivation*	*Scrub*	*OJ*	*DJ*	*OJ (Mainly Teak)*	*DJ (Mainly Teak)*	*G, S+G*	*Mg+ Ma+P*	*Total*
<25	73.6	13.2	0.3	2.7	-	0.3	1.5	8.1	100
25-50	58.2	22.9	3.1	12.7	-	1.3	1.6	0.2	100
50-100	45.8	19.8	7.7	18.4	0.9	5.6	1.6	0.2	100
100-150	24.4	22.7	11.6	28.7	0.9	10.4	1.3	-	100
150-250	23.2	21.4	13.3	30	0.2	11.8	-	-	100
250-350	31.1	17.9	9	32.5	-	9.5	-	-	100
350-550	34.6	20.3	7.1	31	-	6.9	-	-	100
550-750	-	40.6	-	59.4	-	-	-	-	100
>750	-	-	-	100	-	-	-	-	100

(c) Percentage Distribution of Elevation Levels in Each Vegetation Type

Vegetation Type→ *Height (in m)* ↓	*Cultivation*	*Scrub*	*OJ*	*DJ*	*OJ (Mainly Teak)*	*DJ (Mainly Teak)*	*G, S+G*	*Mg+ Ma+P*
<25	32	12.9	0.8	1.9	-	0.2	26.8	95
25-50	16.2	14.3	5	5.9	-	0.6	18	2.1
50-100	24.4	23.5	23.7	16.2	2.9	4.9	35.1	3.2
100-150	6.9	14.4	19.2	13.4	1.4	4.9	15.9	-
150-250	8.8	18.3	29.7	18.9	0.3	7.5	4.3	-
250-350	6.1	7.9	10.4	10.6	-	3.1	-	-
350-550	5.5	7.2	6.6	8.1	-	1.8	-	-
550-750	-	1.6	-	1.7	-	-	-	-
>750	-	-	-	0.3	-	-	-	-
Total	100	100	100			100	100	100

S: Scrub; OJ: Open Jungle; Mg: Mangrove; P: Plantation; G: Grass; DJ: Dense Jungle; Ma: Marsh

The above tables are compiled from analysis of profiles constructed at intervals of 5 minutes latitude

Latitude Across the Study Area

Plantations of palms, casuarina, etc., are seen as narrow stretches along the shore. Mangroves abound in the inter-tidal zone, where the conditions are ideal for their proliferation.

A definite relationship also seems to exist between various slope categories and land cover/land use (Table 2 a-c). Slopes less than 1° have a dominance of cultivation, although other vegetation types also occur. Cultivation shows an expected preference for flat lands and gentle gradients, but some cultivation is practised even on steep slopes. As the gradient increases, say, between 4° and 6°, there is a gradual shift towards scrubs. Scrubs are found in all slope classes in varying proportion.

Table 2

(a) Relation between Vegetation and Slope

Vegetation Type → *Slope (in degrees)* ↓	*Cultivation*	*Scrub*	*OJ*	*OJ (Mainly Teak)*	*DJ*	*DJ (Mainly Teak)*	*G*	*S+G*	*Mg+ Ma+ P*	*Total*
<1	450.8	119.6	22.2	0.8	69.5	26.1	7.3	1.7	20	718.2
1-2	61.9	27	10.3	-	18.7	4.6	0.2	-	2.5	125.2
2-4	61.1	37.6	9.9	1.3	19.5	13.9	0.5	3.6	3.8	151.2
4-6	26.7	32.1	10	1.1	32.5	12.8	0.3	2.2	1.4	119.1
6-10	18.7	41.5	18.8	1.3	64.7	22.1	1.2	0.4	7.1	175.8
10-14	10.4	17.7	12	0.4	47.6	13.2	0.2	0.7	5.1	107.3
14-22	4.7	16.5	12.7	-	39.5	4.8	0.2	0.2	2	80.6
22-30	1.9	20.2	1.9	-	7.4	0.6	-	0.1	-	32.1
>30	0.1	2.6	0.5	-	2	0.6	-	0.1	-	5.9
Total	636.3	314.8	98.3	4.9	301.4	98.7	9.9	9	41.9	1515.4

Note: Values are in km.

(b) Percentage Distribution of Vegetation in Each Slope Class

Vegetation Type → Slope (in degrees) ↓	Cultivation	Scrub	OJ	OJ (Mainly Teak)	DJ	DJ (Mainly Teak)	G	S+G	Mg+ Ma+P	Total
< 1	62.8	16.6	3.1	0.1	9.7	3.6	1	0.2	2.8	100
1-2	49.5	21.6	8.2	-	14.9	3.7	0.2	-	1.9	100
2-4	40.5	24.9	6.5	0.8	12.9	9.2	0.3	2.4	2.5	100
4-6	22.5	27	8.4	0.9	27.4	10.7	0.3	1.9	1.2	100
6-10	10.6	23.6	10.7	0.7	36.8	12.6	0.7	0.2	4	100
10-14	9.6	16.5	11.2	0.4	44.4	12.3	0.2	0.7	4.8	100
14-22	5.7	20.6	15.8	-	49	5.9	0.2	0.2	2.5	100
22-30	5.8	63.1	5.8	-	23.2	1.9	-	0.3	-	100
> 30	1.8	45	7.9	-	34.3	9.7	-	1.4	-	100

(c) Percentage Distribution of Slope Class in Each Vegetation Type

Vegetation Type → Slope (in degrees) ↓	Cultivation	Scrub	OJ	OJ (Mainly Teak)	DJ	DJ (Mainly Teak)	G	S+G	Mg+ Ma+P
< 1	70.7	38	22.6	15.6	23.1	26.4	74.3	19.7	47.7
1-2	9.7	8.6	10.5	-	6.2	4.7	1.5	-	6
2-4	9.6	11.9	10.1	27.1	6.5	14.1	5.1	40	9.1
4-6	4.2	10.2	10.2	21.9	10.8	13	3.1	24.8	3.3
6-10	2.9	13.2	19.2	27.1	21.5	22.4	11.8	4.5	16.9
10-14	1.6	5.6	12.2	8.3	15.8	13.4	2.1	7.9	12.2
14-22	0.7	5.2	13	-	13.1	4.9	2.1	2.2	4.8
22-30	0.3	6.4	2	-	2.5	0.7	-	1.1	-
> 30	-	0.8	0.4	-	1.6	0.7	-	0.9	-
Total	100	100	100	100	100	100	100	100	100

S: Scrub; OJ: Open Jungle; Mg: Mangrove; P: Plantation; G: Grass; DJ: Dense Jungle; Ma: Marsh

The above tables are compiled from analysis of profiles constructed at intervals of 5 minutes latitude across the study area.

Forests occupy steep gradients – 70.7 per cent of the land with slope between 14° and 22° is covered by forests. On gradients steeper than 22°, scrubs become more important than any other category. Steep slopes constitute a fragile terrain with thin soil cover where deforestation leads to a situation where natural regeneration of the original stand is rendered impossible. This leads to the establishment of a degraded form of vegetation, scrubs. These marginal lands require substantial inputs and energy subsidies if they are to be restored to their original status. A small proportion of the land has a slope greater than 30°, where thin soils preclude successful regeneration of vegetation. These areas are susceptible to rapid erosion and must be brought under maximum protection.

Maximum occurrence of open jungle, including teak, is below 1° and 22.3 per cent of the open forests are found here. A substantial amount (19.5%) is found between 6° and 10° and 21.6 per cent between 2° and 6°. Beyond 14°, open forests diminish in importance. This is probably due to a decrease in exploitation of dense forests with decreasing accessibility. It is interesting to note that 23.9 per cent of dense jungles abound on negligible gradients.

Teak in open jungles is important between 2° and 10° of slope. Beyond 14°, no teak is observed in open jungles. The absence of teak above 550 m and beyond 14° slope suggests that such sites are neither favourable nor popular for teak plantations.

Grass generally occurs where the gradient is less than 1°, while mangroves and plantations along the coastline are in areas of imperceptible gradient.

In the process of human intervention, dense jungles get converted to open ones that may further degrade to scrubs and/or grasses to be replaced by cultivation or other land use like transport or settlements.

Deforestation and Degradation of Vegetal Cover

A number of deliberate and unintended courses of action over decades by the people as well as state administration have resulted in a steady decline of forest cover. Settled agriculture was traditionally based on a close association between cultivation and cattle rearing, on the one hand, and between field and forest, on the other. Forests were wooded extensions of cultivated space – used for grazing, hunting, fuelwood, timber, litter, green manure and forest produce to supplement the diet and income of the local inhabitants. Exploitation of forests was based on norms laid down by oral traditions.

Timber extraction has been the main cause of deforestation. Teakwood, highly valued for timber and vessel building, was used locally for centuries and also in Arab lands. It was also used extensively by the Portuguese for ship building at Goa and Vasai.

It appears as though a social regulation of biomass use existed till commercial exploitation began during the colonial period, when the British were bent on increasing land revenue and regarded forests in a negative light. Timber merchants formed a powerful group against effective forest legislation. The absence of any legal framework gave people a free hand and forests were plundered, especially teak stands. Demand for teak was one of the main causes of rapid depletion of the monsoon forests.

Active charcoal production practised by tribal 'Katkaris' and the age-old practice of shifting cultivation are other factors that have led to deforestation. The geographic accident of proximity to tidal ports exporting timber during the late colonial period further contributed to the process.

But none of the above is comparable in effect to that produced by the introduction of rail and road networks on the forest cover of the area (Buchy, 1990; Govt of Bombay, 1950; Govt of Maharashtra, 1964). The pressure intensified with progressive expansion of the transport network in independent India. With the advent of the Konkan Railway, freeways and expressways, such as the Mumbai-Pune Expressway, the trend continues even today.

The rising population led to increasing pressure on forests for which available resources dwindled. Vegetation far inferior to the original forest cover, i.e. scrubs, invaded the degraded forest areas. The delicate balance of relationship between humans, agriculture and forest was disrupted.

Efforts Towards Conservation

The early 1960s witnessed the enforcement of measures for the preservation of forests with customary privileges. The Indian Forest Act VII was formulated in 1878, which was replaced by the Indian Forest Act XVI of 1927 (the Private Forest Acquisition Act was introduced in 1975). Ecological consciousness formed the ideological foundation of the development of state forestry and the creation of the Forest Department in India. The Forest Policy envisaged a concerted effort on the part of the government and people towards planned afforestation with a view to expanding tree lands. Most government forests were classified as 'reserved' or 'protected'. Since access to reserved forests was not permitted, villagers felt no sense of collective responsibility and overexploited the forests that were already degraded. It was then felt that people's involvement and participation could improve the situation substantially; this called for traditions of village solidarity.

In Raigad, the Central and Southern regions are both comparatively less accessible. While the Central region maintained its forest cover, the South has a very limited area under forests. Relative relief is greater in the South and working plan schemes for restoration fewer. This area was formerly a part of the Janjira state with a long history of forest exploitation and shifting cultivation, and marginal imposition of conservation measures.

The steady decline in forest cover in the last century may be attributed to a number of other factors. Unauthorised felling/removal of produce, often with the connivance of forest subordinates, has been an important one. It has been observed that the higher the benefit or rent to the Forest Departments, the greater is the forest loss. Grazing and damage by natural factors like violent cyclones have taken a heavy toll; for example, the 1940 October cyclone uprooted about three million trees in Thane North. Huge demand during the Second World War and the Civil Disobedience Movement in 1930-31 and repeated fires near inhabited areas also played considerable role in deforestation. Of late, there has been a marked decrease in Acacia catechu (khair) trees, primarily due to a heavy demand from the proliferating 'gutka' manufacturing industry.

Concluding Remarks

The environment of North Konkan is suited to support a rich forest cover of tropical moist deciduous or semi-evergreen variety, with glades of pastures. However, with human occupation since millennia, the initial balance in relationship eroded with development, especially after colonial maritime trade from the 16th century A.D. onwards. The natural vegetation was gradually disturbed as per human requirements and commercial demand for specific products.

Mumbai has been the leading hub of all commercial and economic activities, with minor centres like Mahim and Vasai. Prominent creeks penetrating deep into the country increased accessibility till the foothills and tidal rivers like the Savitri, Kundalika and Ulhas strengthened their sea orientation. Later, with the introduction of railways, there was increased exploitation, primarily of teak forests. With rapid strides in industrialisation, coupled with urbanisation, the pressure mounted further. This led to the primary climax forests being replaced by secondary forests of a changed composition.

Within the last couple of decades, forest cover is on the retreat. Scrubs and grasses have replaced forests even at higher elevation and along steep slopes. Mangroves and marshes have declined; for example, in Uran, the built-up area for naval installations has increased remarkably. Maximum deforestation is taking place along rail and road corridors. The effects of the

Konkan Railway and expressways need to be studied in detail to ascertain their impact on the fragile ecosystem.

The Forest Department owns 98 per cent of India's forest cover. The best forests found in the country are represented by reserved forests, which are maintained even at low elevations or gentle gradients in North Konkan. Reclassifying protected and unclassified forests has resulted in an increase in reserved forest area. Vested segments in the forest bureaucracy seek continued expansion of the area under reserved forests. Sometimes, it succumbs to political pressure to regularise encroachments on public land.

The Forest Department, which suffers from lack of adequate resources and manpower, is often helpless in situations like unauthorised felling. More recently, there are instances where collecting firewood is the job of a growing number of women, who chop younger trees since these are relatively easier to cut, but in the bargain, cause considerable damage as regeneration is adversely affected.

It is essential to maintain an average forest cover of 33 per cent. In hilly/mountainous areas, it should increase to 66 per cent for sustenance of the earth's ecosystem. But an area should be considered a forest only if the land has tree canopy exceeding 10 per cent. Not only should the area under tree cover increase, but the quality of forests should also improve, to finally tend towards the original climax vegetation to ensure stability. The Forest Department is trying to improve the condition of open forests and scrublands.

There has been a change in policy as regards forest management over the years. While the first Five-Year Plan encouraged plantations for conservation of soil cover of river catchments, subsequent Plans focused on planting fast-growing and economically important species like teak, matchwood, eucalyptus, bamboo and other timber species. In 1992, participatory forest management was initiated in Maharashtra. The emphasis now is completely on conservation.

It is necessary to evolve a suitable policy framework in order to ensure the required extent of forest cover in the area, which is under a continuous threat of urban and industrial expansion. This would not only save the forests, but also provide recreational space to urban dwellers.

Acknowledgement

The author wishes to thank Prof. B. Arunachalam, former head, Department of Geography, University of Mumbai, for guidance and inspiration.

References

Buchy, M. (1990): 'Colonial Forest Exploitation in the Western Ghats of India: A Case Study of North Kanara District', Pondy Paper in Social Sciences, French Institute, Pondicherry.

Govt of Bombay (1950): *The Bombay Forest Manual*, Vol. III, Mumbai.

Govt of Maharashtra (1964): *Kolaba District Gazetteer*, Mumbai.

Meher-Homji, V.M. (1987): 'Vegetation Types of Peninsular India in Relation to Environmental Conditions', in V.S. Datye et al. (eds), *Explorations in the Tropics*, University of Pune, 3-19.

Srivastava, S. (1995): 'The Phytogeography of North Konkan (with Special Reference to Forest Cover)', Unpublished Ph.D. thesis, Department of Geography, University of Mumbai.

15

Distribution of Agricultural Land among Social Groups in Western Maharashtra

A.S. Jadhav

Introduction

Most of the freedom fighters of India ignored social and economic issues during their struggle for independence, anticipating that such problems would be resolved on securing political freedom. However, even after 50 years of independence, economic disparities continue to exist. Such disparities are very pronounced in land ownership.

Agriculture being a major economic activity in the country, land plays an important role and is the principal means of production in rural India (Mandavdhare, 1989). It continues as an instrument of social power (C.P.I., 1949; Planning Commission, 1949; Bose, 1972) Access to land determines the status of a social group in the economic and social hierarchy. Among such groups, those that hold land are mostly formed of dominant castes and, hence, inequality in land distribution is not only an economic, but also a social problem (Gupta, 1986).

Generally, the people belonging to the higher castes own a major share of good quality and productive land, while the low, backward castes or tribes belong to the community of marginal farmers and landless labourers (Malhotra et al., 1976; Bhagabati, 1990). Landless labourers are exploited at the hands of the higher castes. The agrarian problem thus cannot be studied in isolation, as it is complex in nature and interlinked with the social hierarchy; it cannot be appreciated without understanding the social hierarchy based on caste and class formations, which is revealed by structured access to the means of production. This study was undertaken with a view to investigate the nature of the relationship between social groups and land ownership at the village level in a part of Western Maharashtra, selected on the basis of empirical evidence. It also attempts to analyse the distributional pattern of average size, disparity and different size classes of landholdings in parts of Western Maharashtra.

Objectives of the Study

The problem is directly linked to selected socioeconomic aspects of the various social groups. The specific objectives of the study are:

1. To investigate the imbalances in agricultural landholdings in different social groups and find out the nature of relationship between the social groups and land ownership;
2. To examine the distributional pattern of landholdings among the scheduled castes/tribes and other social groups;
3. To appreciate who actually controls the landholdings in villages and see how the land is distributed among social groups;
4. To explore the nature of irrigated landholdings and the groups which own most of the irrigated lands; and
5. To identify groups which possess small and marginal landholdings.

Research Questions

No hypothesis has been formulated as the relationship of social groups and land is very complex. But an attempt is made to seek answers to some of the questions regarding the land and group relationship. Some of the important research questions investigated during the course of this study are:

(1) Does the social hierarchy in rural India reflect the economic hierarchy?
(2) Does caste per se generate disparity in ownership of land?
(3) Does caste as a factor operate at the time of settlement of the primeval groups on pioneer lands leading to the occupancy of the best lands of the village by the middle ranking castes, which eventually emerge as the dominant castes?
(4) Can one infer from the above that terrain and agroclimatic factors play a significant role in determining inter-caste disparity in land ownership?
(5) Is it true that the development of agricultural infrastructure, such as irrigation, accentuates further the inequality in land and assets among the caste groups?

Study Area

The study area forms a part of the state of Maharashtra. It comprises five villages from four districts belonging to the Southwestern parts of the plateau. The details of the villages selected for the purpose are as follows:

1. Hamidwada – taluka Kagal, district Kolhapur;
2. Savantpur – taluka Tasgaon, district Sangli;
3. Vajrachaunde – taluka Tasgaon, district Sangli;
4. Udanwadi – taluka Sangola, district Solapur; and
5. Wavarhire – taluka Dahiwadi (Man), district Satara.

Data Base

Data on the relevant socioeconomic aspects were collected from both primary and secondary sources on a selective basis.

The primary data were generated through specially designed questionnaires. Villagers selected by sampling filled in these questionnaires. The information so collected related to aspects such as the caste/tribe they belong to, size of landholding, general land use, land under cultivation, irrigated land, cropping pattern, agricultural implements owned and production of principal crops.

Secondary data were obtained from official documents like the District Socioeconomic Abstracts and Statistical Review, District Gazetteers and Census records. Data were also collected from unpublished sources like the official records of the Talathi (village accountant) and Gram Sevak (village official) on the total populations of villages, their caste-wise break-up, geographical and irrigated areas of villages, land-use patterns and principal crops.

Methodology

In order to understand the relationship between different social groups and landholding patterns in the study area, it was thought that sampling would be a suitable method for data collection. Only five villages from four districts in Western Maharashtra, Kolhapur, Sangli, Satara and Solapur, have been selected due to constraints of time. The selection is arbitrary, based on convenience of fieldwork. The social groups have been carefully selected by purposive sampling to provide fair representation to all castes/tribes from each village. About five to ten households from each caste/tribe have been selected by employing the random sampling technique. About 60 households have thus been selected from each village, totalling 300 households in all.

Four major groups are found in each village - scheduled castes, nomadic tribes, other backward castes and castes falling under the open category. Castes/tribes which are notified by the governments of India and Maharashtra are included in the respective categories and all others are treated as belonging to the open category.

Data collected through observations and questionnaires have been tabulated groupwise for each of the villages.

Households have also been grouped according to farm size as marginal (with 0.01 to 0.99 ha), small (with 1 to 2.99 ha), semi-medium (3 to 4.99 ha), medium (5 to 9.99 ha) and large (with more than 10 ha). The relative status has been derived using percentages to facilitate comparison among the groups. The results are presented in the form of a simple table.

Use of highly sophisticated statistical and quantitative techniques generally employed in geographical research has been avoided. This has been done largely due to the fact that the relationship between land and social groups is not linear and thus not easily amenable to a priori hypotheses. Qualitative judgements have been preferred to readymade quantitative tools because of the complex nature of the relationships involved.

Comparative Status of Various Social Groups

The land distribution pattern among the groups is complex and interlinked to the problems of social hierarchy. To understand the nature of the society, in the Indian context, it becomes imperative to study the distribution of land among the different social groups. An attempt has been made to understand the relationship between caste/tribe groups and land ownership in five villages.

The five villages under study display diversity in their caste/tribe composition, terrain type, soil, rainfall, agricultural inputs and level of development over time. Some villages have attained a higher level of technological development, while others are still lagging behind.

There are in all 30 caste/tribe groups in the study area. Out of these, nine castes in the open category hold 67.49 per cent, scheduled castes hold 13.11 per cent, nomadic tribes hold 10.06 per cent, while other backward castes hold 9.34 per cent of the land. The dominance of upper castes is evident from their occupying over two-thirds of the land. Scheduled castes and nomadic tribes compete for second place and other backward castes for third place. One also finds variation in the size of the landholding among the various groups (Table 1).

Table 1
Landholders and Landholdings in Five Villages

S. No.	*Village*	*No. of Caste/ Tribe Groups*	*No. of Land-holding Families*	*Area of Land-holding (in ha.)*	*Average Land-holding (in ha.)*	*No. of Landless Families*
1.	Hamidwada	18	47 (78.33)	50.76	1.08	13 (41.66)
2.	Savantpur	18	36 (60.00)	58.63	1.62	24 (40.0)
3.	Udanwadi	14	45 (75.00)	98.46	2.19	15 (25.0)
4.	Vajrachaunde	9	37 (61.66)	157.85	4.26	23 (38.33)
5.	Wavarhire	15	59 (89.33)	83.39	1.41	1. (1.67)
	Total	30	224 (74.66)	449.09	2.00	76 (25.33)

Total No. of Social Groups (Castes/Tribes): 30

Figures in parenthesis are percentages

Source: Compiled by the author

Most SC/NT families do not hold any land (Table 2). Actually, in the entire study area, 75 per cent of the households are land owners and 25 per cent are landless. Most of the landless labourers hail from backward castes such as Mang, Mahars and Chamars. Social hierarchy thus is intertwined with economic hierarchy and upper caste people are mostly found economically better off than the lower caste people. Maratha is the most dominant caste in the study area; their share in the population is 17.94 per cent, 51.80 per cent, and 49.37 per cent in Hamidwada, Savantpur and Vajrachaunde, respectively, but they hold the major chunk of land in these villages, i.e. 35.82 per cent, 92.44 per cent and 80.36 per cent, respectively. The land held by this caste is almost the twice the proportion of its population. Udanwadi is dominated by a nomadic tribe (the

Table 2
Status of Social Groups with Respect to Size of Landholding

S. No.	Village	Caste Categories	Population (%)	Land (%)	Irrigated Land (%)	Unirrigated Land (%)	Marginal Land-holders (%)	Small Land-holders (%)	Small Medium L.H. (%)	Medium L.H. (%)	Large L.H. (%)
1.	Hamidwada	Open	37.79	68.16	73.77	64.85	22.72	70.81	100	-	-
		SC	28.53	8.27	-	14.06	36.36	4.16	-	-	-
		OBC	33.65	23.56	26.24	21.08	40.70	24.99	-	-	-
		NT	-	-	-	-	-	-	-	-	-
2.	Savantpur	Open	64.11	93.45	96.90	83.02	58.32	92.30	100	100	100
		SC	13.10	6.10	0.96	11.58	24.99	7.69	-	-	-
		OBC	12.71	1.17	2.12	0.57	16.66	-	-	-	-
		NT	10.80	-	-	-	-	-	-	-	-
3.	Udanwadi	Open	39.12	55.09	66.12	53.02	23.52	23.52	66.66	75.00	100
		SC	31.47	10.72	1.83	13.66	47.05	23.52	-	-	-
		OBC	9.71	4.10	5.89	1.34	17.64	11.76	-	-	-
		NT	19.70	30.10	26.26	31.98	11.76	41.17	33.33	25.00	-

Contd...

Contd...

4.	Vajrachaunde	Open	49.37	80.36	87.93	70.01	77.77	56.25	100	100	100
		SC	24.68	15.51	9.82	23.23	11.11	37.50	–	–	–
		OBC	8.63	2.53	2.19	2.99	–	6.25	–	–	–
		NT	17.49	1.58	–	3.73	11.12	–	–	–	–
5.	Wavarhire	Open	36.71	40.06	62.91	31.72	30.30	23.53	42.86	100	–
		SC	34.52	25.93	11.44	33.98	39.39	29.41	28.57	–	–
		OBC	16.44	15.68	15.42	17.48	16.15	23.53	14.28	–	–
		N.T.	12.33	18.02	10.23	16.71	16.15	23.53	14.28	–	–

Note: Maratha is the dominant caste on irrigated land (27.89, 96.71, 87.93) and on unirrigated land (42.19, 82.22, 70.01) in Hamidwada, Savantpur and Vajrachaunde villages, respectively. Lingayat Reddy is the dominant caste on irrigated land (33.01) and Dhangar is the dominant caste on unirrigated land (31.31) in Udanwadi village. Brahmins are the dominant caste in terms of land (32.60) and unirrigated land (28.26) in Wavarhire village.

Source: Compiled by the author.

Dhangar). The tribe's share of population is 18.23 per cent and it holds 29.69 per cent of the land. This domination is only because the terrain in this area is hilly and most of the tribals domesticate sheep. Their agriculture is mostly dependent on rainfall and consists of subsistence farming. Brahmins dominate Wavarhire. Their population is 10.68 per cent of the total, but they possess 25.27 per cent land.

In four out of the five villages, land is controlled by people belonging to the open categories and in only one village do nomadic tribes control the land. In the case of irrigation, it is found that most of the irrigated land is owned by Marathas and other upper caste people. The condition of the scheduled caste farmers is very pathetic. Data shows that about 1 per cent land of the SCs is under irrigation as against 27-96 per cent of that belonging to the open categories. From field observations, it was found that among the nomadic tribes, the Dhangars are the most dominant in all the villages. Udanwadi has the largest share of landholdings owned by Dhangars, i.e. 29.69 per cent of the total village land.

Concluding Remarks

It is evident from the statistics related to size of landholdings that there are a few large-scale landholders in the study area and they are from the Maratha, Brahmin and Lingayat castes. The majority of the scheduled caste people have either marginal or small landholdings. It is observed that their landholdings are not economically viable. Because of the small and marginal size of their landholdings, SC/NT people work on the lands of the other castes.

Enquiry in the study area reveals that upper caste farmers own the lands surrounding the villages and these lands are fertile and of better quality, whereas most of the *Barad/Malran* (barren, less fertile land, located away from the village) are owned by the scheduled castes.

While investigating loan and other government facilities availed of by the different social groups, it is noticed that most of the facilities are cornered by the high caste people; the lower caste people are not even aware of the facilities provided by the government and other agencies. There are some high caste farmers who have not availed of any farm facilities and have no animals or implements. They are well off and have leased out their lands to other tenants on a sharing or contract basis. Such cases are found in most of the villages. Most of these high caste landholders are professionals, such as doctors, lawyers, service holders or businessmen.

Socially and economically backward groups like scheduled castes and nomadic tribes mostly do not possess agricultural implements and, most of the time, they have to depend upon the high caste people. Sophisticated and mechanised equipment like tillers or tractors are owned by the large landholders who are high caste people. Even in the case of domestic animals, the picture is the same.

Disparity in landholdings leads to economic and social disparity. A large section of the population cannot avail of other facilities owing to their poverty. It can, therefore, be concluded that one of the principal causes of socioeconomic disparity in rural areas is embedded in the land distribution pattern.

References

Bhagabati, A.K. (1990): 'Social Structure and Agricultural Performance in Dolabri Irrigated Area of Sonitpur District in Assam', *Geographical Review of India*, 52(4), 71-9.

Bose, N.K. (1972): 'Experiences in Cultural Enquiries', *Man in India*, 52(3), 201-11.

Communist Party of India (1949): *On the Agrarian Question in India*, Peoples' Publishing House, Mumbai.

Gupta, M.S. (1986): 'Distribution Patterns of Land Holding in Madhya Pradesh', *The Deccan Geographer*, 24(1&2), 29-38.

Mandavdhare, S.M. (1989): *Caste and Land Relations in India*, Uppal Publishing House, New Delhi.

Malhotra, S.P., Bharara, L.P., and Sen, M.L.A. (1976): 'Land Utilization and Agricultural Situations in Bikaner', *The Deccan Geographer*, 14 (1), 1-9.

Planning Commission (1949): Report of Agrarian Reforms Committee, Govt of India, New Delhi.

16

Functional Change and Population Displacement within Core Areas in Greater Mumbai

Heinz Nissel

Background

During the 1980s, the rate of urbanisation in India slowed down while the increase in the urban population was enormous – an increase of 57 million persons, adding up to 217 million urban residents. The paradox still holds true that India is experiencing rapid urban growth and simultaneously slow urbanisation. Urbanisation increased from 23.3 per cent to 25.7 per cent only, meaning three out of four Indians still live in the countryside. Census reports on urban population by size class show that from 1901 onwards till date, due to population increase, there is a shift from small and medium towns to Class I cities. Large cities accounted for 65 per cent of the total urban population in 1991 against just 26 per cent in 1901. In accordance with almost all developing countries, the bigger the city, the bigger its pull. This means that inside the Class I category, India's 23 'million cities' are gaining constantly more than other cities; almost one-third of all urban residents today live there. While one can argue that this was a process of metropolitanisation that has operated since the 1950s onwards, since the 1980s, this accelerated development has contributed to a new dimension, i.e. the growth of mega cities. More than 50 per cent of the total inhabitants of million cities are now living in just the four mega cities, Mumbai, Kolkata, Delhi and Chennai, comprising about 37 million inhabitants in 1991. While mega cities are defined functionally by their position as nodes, in a globalised world economy, one can argue they are statistically seen as urban agglomerations holding more than five million inhabitants.

During the last decade, the Mumbai urban agglomeration gained not less than 4.3 million in population, thereby surpassing more than 12.5 million inhabitants and taking the leading position in the country. This tremendous population increase can be seen as a reflection of the ever increasing power of the city in India and its leading position in the national networks of trade, commerce and communication. This paper argues that functional change as well as population displacement within core areas of Mumbai are closely connected with the processes of metropolitanisation and mega city development.

Creation and Perpetuation of Two Urban Core Areas

The processes of functional and population change started in the early 1960s. In previous articles (Nissel, 1977; 1978), the author has developed a zoning model for Mumbai, explaining the sharp dichotomy between the Fort Area and the Bazaar Zone. Fort Area on the Southern tip of the Island City is the site of the 'old town', the legacy of the British since the early 18th century and at present the 'Westernised city'. As far as city administration is concerned, this area corresponds to Ward A.

North of it, the Bazaar Zone came into being during the early 19th century, originally called the 'native town', and developed in several historical phases roughly corresponding to the present Wards B and C. Sections 25 and 26 of Ward E and, at a later stage, section 18 and some circles from Ward D, became a part of it.

Originally, the 'old town' contained different population groups from several regions of India, Arabia and Britain. After a big fire destroyed nearly half of it in 1803, the poorer sections were resettled in the 'native town'. The economic uprise in the mid-19th century, in the form of the 'cotton boom' or 'share mania', brought about, within a few years, the present infrastructural layout that went to make the backbone of the metropolis today, comprising the big public buildings and wide avenues. While the inner core of the Fort Area developed into the city's central business district with its concentration of banks, insurance companies and business headquarters, the upper and middle class, mainly made up by the British and Parsi citizens, shifted their residential quarters to Colaba and the Malabar and Cumballa hills. Population displacement around the Central Business District (CBD) was thus a part of the city expansion being an outcome of the spatial spread of tertiary activities and a resultant increase in urban land values. On the other hand, several living quarters of the 'native town', already a century old, had developed into market areas and went to make the Bazaar Zone, the most important base for wholesale and retail trade all over India. While shops and hawkers still dominate the physiognomic appearance of the streets, this is in fact a multifunctional core area of Greater Mumbai. Services, residential function and small-scale 'backyard' industrial production all play a vital role, existing side-by-side in a mixed land-use pattern, where living and working are closely interwoven.

Functional changes and heavy population losses, in spite of the tremendous increase in the overall population of the mega city, continue within the core area of Mumbai.

Population Losses: Facts and Figures

In the Fort Area (Ward A), at a first glance, one observes a slight increase in population during 1971-91. Quite different is the situation in the Bazaar Zone. In 1961, most sections were overcrowded to such an extent that urban researchers in a worldwide comparison paralleled them with the population densities of Hong Kong. Bhuleshwar and Kumbharwada, for instance, were burdened with more than 400 persons per ha, while all the other sections were holding not less than 200 persons per ha. Looking at the Census figures for 1971, one can make out that Khara Talao and Fort North, constituting the very heart of trading activities, lost over 10 per cent. All the other sections lost people, but not so much as to lower the density pattern to a considerable extent. After 1971, further decline of population gained momentum; within 20 years, Ward B lost 32.8 per cent of its inhabitants and Ward C, 36.9 per cent. Speaking in absolute numbers, Ward B declined from 175,000 to 118,00 inhabitants, and Ward C, from 312,000 to 197,000 inhabitants. That is to say, within a comparatively short time, the population within both the wards declined by around 173,000 persons or roughly one-third. Sections in the adjoining areas of Wards D and E, which belong historically, functionally and in the population composition to a large degree to the Bazaar Zone of Girgaum, Khetwadi and Chaupatty within Ward D, Kamathipura and 2nd Nagpada from Ward E show the same pattern of population decline of between 20-35 per cent in the same time period (Table 1).

If we compare these figures with the development of Greater Mumbai, we find that the combined population share of Wards A, B and C have shrunk from 17.1 per cent (1961) to merely 5.1 per cent (1991) of the total, but they still hold about one-third of the establishments as well as tertiary workers and one-quarter of all shops. It seems, at a first glance, that both core areas of trade and commerce still play a vital role at the cost of the residential function. However, at the same time, the

Table 1
Population Losses in Core Areas of the Island City, 1971-91

		Per Cent
More than Half of the Population		
Section (4)	Fort North	–50.2
Section (13)	Market	–57.3
More than One-third of the Population		
Section (3)	Forth South	–35.3
Section (6)	Mandvi	–35.8
Section (7)	Chakala	–39.0
Section (8)	Umerkhadi	–33.0
Section (10)	Khara Talao	–35.2
Section (12)	Bhuleshwar	–34.0
Section (14)	Dhobi Talao	–39.1
Section (19)	Chaupatty	–33.5
Around One Quarter of the Population		
Section (5)	Esplanade	–21.6
Section (9)	Dongri	–27.7
Section (11)	Kumbharwada	–27.9
Section (15)	Fanaswadi	–29.0
Section (16)	Khetwadi	–25.6
Section (18)	Girgaum	–29.1
Section (25)	2nd Nagpada	–21.9
Section (26)	Kamathipura	–20.8

population composition between the Island City and suburbs changed even more dramatically with the huge population increase in the latter. In 1961, the relation between the inhabitants of the Island City and those of the suburbs showed a ratio of 2:1. In 1971, it had changed to 1:1, in 1981, 1:2, and in 1991, to 1:3. In spite of the population expansion in the suburbs and in Navi Mumbai on the other side of the Thane Creek, in 1991, just one-quarter of Greater Mumbai's inhabitants were found in the Island City. But the area still holds two-thirds of all tertiary activities, 60 per cent of shops and nearly three out of four commercial establishments. This is a strong indication of the ongoing commercialisation of the old urban cores, while more and more citizens are forced to seek refuge in further outlying areas, leading to increasing segregation between living and working places and aggravating the problem of daily commuting to an unmanageable level.

This heavy population displacement has had a considerable impact on the population density pattern. Fort South, the inner core of 'big business', is now only one section out of 38 in the Island City, holding less than 20 inhabitants per ha. On the other hand, there is no section left containing more than 400 persons per ha and just four of them still having more than 200 inhabitants per ha: These are Bhuleshwar, Umerkhadi, Kumbharwada and Kamathipura, forming parts of the Bazaar Zone, where the commercial functions have not displaced the living quarters of the urban poor until now. All other sections have lowered their density rates below 200 persons per ha, representing a similar pattern as of Byculla, Dadar or Naigaum today. Dongri has even fallen below the 100 per ha mark.

Although the core areas experience by far the highest intensity of population decline, it would be wrong to overlook the fact that during the last two decades a number of other sections in Wards E, F and G, mainly forming the industrial belt of the Island City, i.e. textile mills and workers' quarters (*chawls*), have shown considerable population loss as well. In Parel and Matunga, the population has decreased by 14 per cent, in Naigaum by 17 per cent and in Prabhadevi, by 6 per cent. The most important section of Ward E, Byculla, has lost 11 per cent, while Tardeo and Tadwadi have lost slightly more. But in many of these areas (Byculla, Matunga and Mahim) one not only finds workers' quarters housing the poor population, but also well-laid out quarters for middle income groups built between the two World Wars. This indicates that better-off families, with members holding white-collar jobs, have been able, in the last 20 years, to shift on their own to the suburbs in search of better residential quarters, while workers, especially those working in the textile mills, were forced to leave their residential quarters in these areas. The real big increase in population within the Island City occurs in the Northernmost sections of Sion and Mahim, which had small houses within coconut palm groves and salt pans even around 1970. Both together added around 3,40,000 persons within the last two decades.

From the foregoing discussion, one finds that there still exists a peak in those four sections of the Bazaar Zone already mentioned, with population densities of more than 200 persons per ha. They are surrounded by sections holding between 100 and 200 persons per ha and these continue as a belt comprising sections that lie towards the North. Most of the other sections show densities of between 40 and 100 persons per ha indicating a certain adjustment during the last decade. Only four sections are found with lower densities, viz. Fort South, Esplanade and Middle and Lower Colaba of Ward A and Dongri, a part of the harbour side of the Bazaar Zone.

Population Decline: Causes and Effects

Fort Area

Development in the Fort Area can be said to have been initiated by national and international factors that actually provided the physical setting for a mega city. There is the continuous pressure of Mumbai's economic drive on the citizens still living in Fort Area, who have become extremely vulnerable to displacement. Within Fort South, the old buildings, a legacy of the British, still exist, but their use as residences faded away decades ago. Currently they house a number of auxiliary city functions, such as photocopy shops, small restaurants, etc. On the other hand, the old parts of Fort North have still partially preserved a residential setting. However, many of the old buildings have been demolished and half of the population displaced during 1971-91. Tertiary functions and economically stronger population groups have come in and replaced the former ones. Naturally, there is no space left for weaker groups of the urban society, still dependent on low-rent housing.

In spite of such economic processes and changes, the oldest parts of the metropolis are not in the forefront of CBD expansion. With the development of the Back Bay, there has been a shift of the highest ranking tertiary and quaternary functions – headquarters of banks, industrial firms, commercial agencies, consulates, airline offices, etc.; skyscrapers now thrive on this most expensive piece of land. The residential function is virtually non-existent. It might be interesting to note that the original intention of the Nariman Point reclamation was the construction of middle-class apartments, which never came into being. The second line of CBD expansion is directed to the South in Colaba. A large number of hotels, shops and restaurants, especially those catering to foreign tourists, put an immense pressure on the old residents. Traditional cottages, many of them owned by Parsi families, have been replaced by costly high-rise apartments. 'Citizens' disappear fast, the 'nouveau riche' come in. As Colaba was already a place with upper middle class status, the process marked the replacement of an 'old elite' with a 'new elite', based on income, with traditional elements of social segregation becoming a thing of the past in areas like Cuff Parade.

City expansion does not even stop at the precincts of the urban elite quarters. Through the transformation of Mumbai into a mega city, there is a constant demand for office space in high-ranking areas. Therefore, even in the upper class areas, such as the Malabar and Cumballa Hills, residents are threatened by the intrusion of consulates, management firms, foreign business houses, etc., able to afford higher rents. Through globalisation, India's large number of talented software specialists is used by an increasing number of business firms, joint ventures, etc. Since the 1990s, it has become common to transfer the book-keeping of important companies to India, leading to cost reduction of up to 90 per cent. For instance, leading airlines like British Airways, Lufthansa,

Swissair and Austrian Airlines have outsourced their office work to Indian cities, including Mumbai. This is a good example of Mumbai becoming a node in the global economic network.

Bazaar Area

It is the functional mix of living and working that defines the second urban core of Mumbai with bazaar streets, markets and hawkers as dominant elements. The economic scene resembles a quite complicated, but fascinating, pattern of buying and selling, retail and wholesale trade, backyard production units and daily living, more as a rule than as an exception, appearing at times under one roof in one large family, where economic functions might be divided among the members. In spite of this intensive use of space for economic purposes and heavy population losses within the last two decades, residential densities are still high. As mentioned earlier, social and cultural norms, values and forms characterise the organisation of the Bazaar Zone to a larger degree than economic necessities. Too many shops display similar goods, indicating poor business for many items. In the 1970s, I had estimated that all the wholesale and retail dealers combined to serve one merchant, representing 14 inhabitants or three families, a strong hint at the somewhat parasitic character of jobs as well as disguised unemployment. As there is such a massive population loss in so many sections, it is clear that many of the small shop owners, traders and hawkers, who must have lost their customers, could not make a living and were forced to move to the suburbs. However, some old principles still work: Religious faith defines not only the way of life, but also what goods are offered to cater to the daily needs of inhabitants. Most professions are still not acquired, but inherited. Various areas of economic life are monopolised by certain communities, e.g., Gujaratis work as textile traders and jewellers, Marwaris as moneylenders, Tamilians and Keralites as clerks or waiters in restaurants, Punjabis as transport operators, all occurring with a regularity.

It is important to note that most of these members of the working population live in the same houses where they work, giving rise to working neighbourhoods. At a glance, these living quarters seem to be chaotic or amorphous, but if one looks closer, one observes a rather complicated, but well organised, mosaic of neighbourhoods. A high degree of community spirit prevails in these units even today. In contrast to the CBD and upper class areas, they are not differentiated along income, occupation, education or prestige lines, but according to religion and language of the inhabitants. The Muslim quarters are strongly segregated from those of the Hindus, whose districts are divided by caste or even sub-caste. Slum areas in the Bazaar Zone are not meant as quarters of especially low social status, but those with extremely poor living conditions as far as age of the housing stock, population density, etc., are concerned. Kamathipura is a good example.

Putting the commercial functions of the Bazaar Zone in a general framework, I feel that they are mostly concentrated in the South and the centre of the area, i.e. in the sections of Market, Chakala and Bhuleshwar. Moving North, the residential character gets stronger. While further North, there still exists a mixture with small-scale industries, many of them disappeared during the last decade. Most of the shops still show traditional elements, often 2 or 3 metres wide, open during the day and locked with wooden doors at night. Handicrafts and vocations have moved into sidestreets and backyards. Traditional occupations are threatened by the sale of cheap articles of industrial mass production, many coming from East Asia. Workers in traditional occupations or household industries are forced to shift from the production of items to merely repairing them.

In contrast, businesses with large profits and higher prestige assimilate the external and internal appearance of the shops in the Fort Area and Colaba, with huge glass windows to exhibit the products, fixed prices and air-conditioned showrooms. These changes are typical for the establishments of

Figure 1

Bazar Area of Mumbai (Ward C) Spatial Concentration and Differentiation of the Wholesale and Intermediate Trade

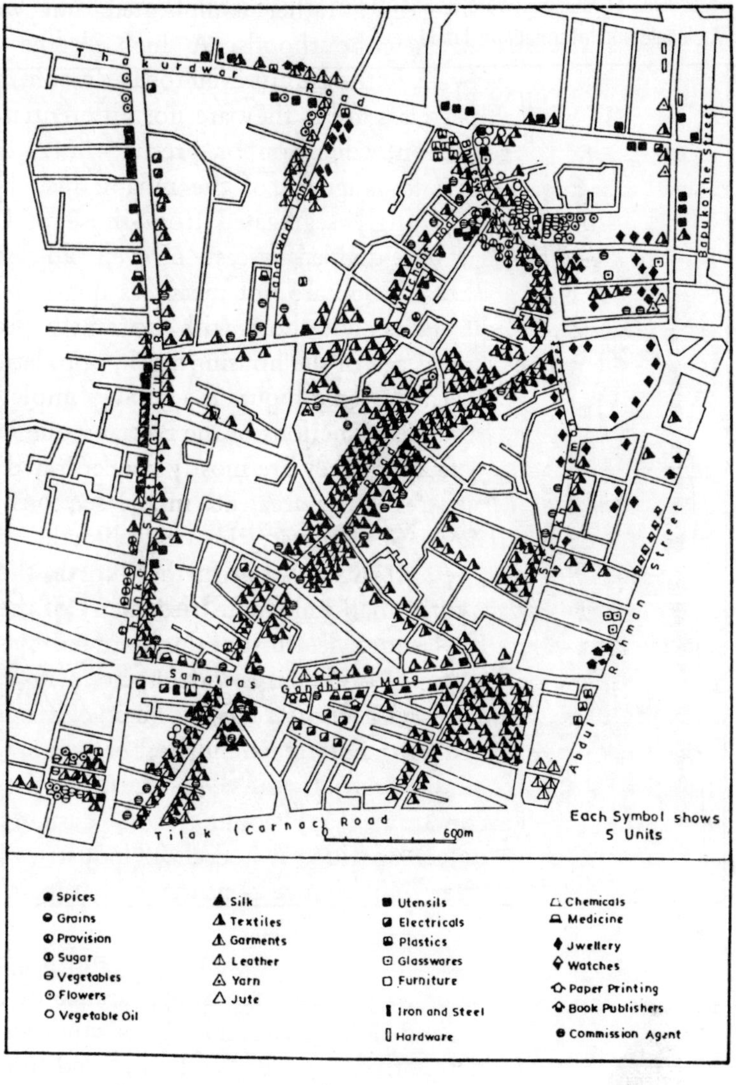

Figure 2

Bazar Area of Mumbai (Ward B) Spatial Concentration and Differentiation of the Wholesale and Intermediate Trade

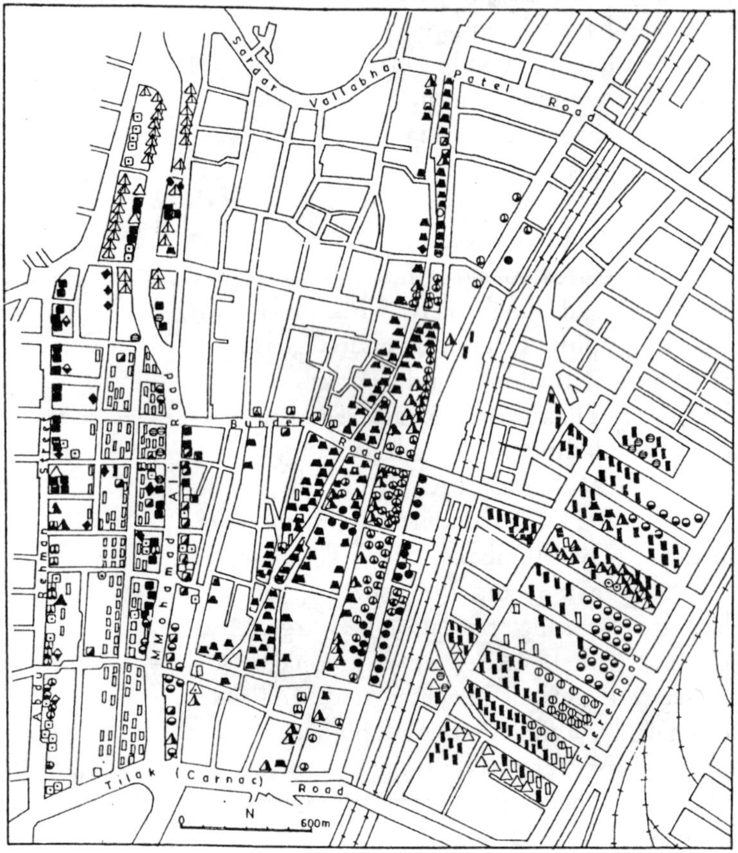

jewellers and goldsmiths, and those selling chinaware, glass, ceramics, electrical appliances, sportswear, specialised textile units, etc. Within the framework of this article, it is not possible to show all the differentiations of the markets, bazaar streets and their respective living areas (Figures 1 and 2). Especially in the Eastern parts along the harbourside, a number of changes have taken place.

The largest textile markets, viz. Mulji Jetha and Mangaldas, are found on the Southern rim of the Bazaar Zone, surrounded by countless retail shops, most of them in the hands of Hindu Gujaratis. The same holds true for the whole length of Kalbadevi Road, Sheikh Memon Street and the Mumbadevi area, which hold Mumbai's most prominent jewellery and gold shops; two decades ago, these shops enclosing Mumbadevi temple were dealing solely with silver ornaments, thereby indicating uplift. Chemicals and electrical stores are concentrated along Samaldas Gandhi Marg and Lohar Chawl, paper printing at Thakurdwar Road, vegetable and flower markets around Bhuleshwar Street and, finally, building materials and furniture at Mohammed Ali Road. The grain and spice markets in Mandvi and Dongri and iron and steel trading have now been shifted to Navi Mumbai.

Between 1961 and 1971, several areas of the Bazaar Zone were already losing population to a large degree; Market lost 18 per cent, Chakala, 14 per cent, Kumbharwhada, 22 per cent. All of them are sections in which trade and small-scale industries dominated. On the other hand, Bhuleshwar and Kumbharwada lost only 4 per cent each. In the next two decades Market lost a staggering 57 per cent from the 1971 population level, Chakala, hardly less than 40 per cent. The same happened to Dhobi Talao, while adjoining Mandvi faced a decline of 36 per cent. The shift of so many wholesale markets as well as retailers to Navi Mumbai – from iron and steel to grain and spices – destroyed the vital economic base of the core.

Population displacement gained speed in sections of the Bazaar Zone where the living function was not overwhelmed by trading activities. The very heart of the zone, Bhuleshwar, lost 34 per cent and Girgaum and Fanaswadi, around 30 per cent. But these areas simultaneously acted as anchoring points for the local economy and provided the local inhabitants with a vital source of income. Segregation along religious lines still occurs mainly along a boundary between Kamathipura and 2nd Nagpada, continuing Southwards between Kumbharwada and Khara Talao, then splitting the circles of Bhuleshwar into an Eastern and Western half, finally moving into Market from Chakala. Along the Western side, the Hindus prevail, to the east, the Muslims. In Chakala, Muslims had a 77 per cent majority in 1961, while Hindus accounted for only 17 per cent. But in 1931, the number of Hindus was twice as large. Hindus were to be found in Kumbharwada in 1961 in an overwhelming majority of 88 per cent, the same was true for the Muslims within the neighbouring section of Khara Talao (85%).

All these raise a question: Is there a typical distribution of working and living patterns along communal lines within the Bazaar Area because it is economically efficient? To answer that question, it would be necessary to use an integrated approach to show how the different spatial levels – local/urban/regional/national/global – are interdependent. One must not forget that for millions of common people, these old core areas of metropolitan agglomerations worldwide are important to make a living, to survive in spite of the brutal socioeconomic divisions that characterise our present urban societies by and large. Further, functions in this economic pocket are well supplemented by cheap housing stock and low rents. Authorities must be very careful while improving housing or traffic conditions, keeping in mind the day to day fights for survival of the citizens living and working there.

The living conditions in many of the sections of the Bazaar Zone are bad – with old, dilapidated houses, pollution, noise and traffic jams, lack of parks and open space. It might not be wrong to think that many of the tradesmen and their families who got used to living in the area finally left for the suburbs to settle in a better environment.

The factors of population displacement in the Fort Area and the Bazaar Zone are interrelated with the developments mentioned above. These are: (i) problems of physical setting (buildings and sites, infrastructure); (ii) economic and social uplift of certain groups of inhabitants; (iii) population displacement through economic pressures (i.e. shifting of wholesale markets to Navi Mumbai); and (iv) problems of increasing communal tension.

It could be possible that one of these factors is responsible for most of the changes. It might also be true that all these factors are working together.

Concluding Remarks

There are highly conflicting viewpoints regarding how a city centre should be put to proper use. Many experts would prefer to go with globalisation and do away with outdated modes of living and working. New ways of dealing with planning problems, as learnt nowadays in Europe, can be thought of, i.e. going to forms of grass-root planning, involving serious discussions with and true participation of the inhabitants of the local working neighbourhoods. This in no way is an easy task, but it may pay off in the long run. It needs to be remembered that creating or preserving housing for lower income groups while simultaneously guaranteeing a diversity of functional uses and protecting cultural inheritance is not an easy task.

Note

Some ideas and results of this paper were discussed first at a conference on 'Working and Living in Cities' in November 1994, organised by Max Mueller Bhavan, Mumbai, and held in the University of Bombay.

References

Nissel, H. (1977): 'Greater Bombay: An Explanation of Structures, Functions and Development', *Bombay Geographical Magazine*, XXV (1), 1-24.

Nissel, H. (1978): 'Determination of Social Areas by Using Multivariate Techniques: The Case of Greater Bombay', *Bombay Geographical Magazine*, XXVI (1), 1-31.

17

Manufacturing Workers in the Informal Sector: Typology and Organisation of *Jari* Workers in Mumbai

Smita Gandhi and S. Banerjee-Guha

Introduction

The continuing existence of traditional activities in Third World urban economies is an expression of the coexistence of pre-industrial forms of production with modern manufacturing. Craft activities among these, especially in India and other Asian countries, have a long tradition and strong presence (King, 1979), unlike in Latin American countries such as Brazil and Peru, where they are on the decline (Kowarick, 1979; Scott, 1979). During phases of urban and industrial growth in many countries, these archaic forms of production in cities have got linked with the dominant capitalist mode of production and undergone transformation to finally get dissolved (McGee, 1979).

In Mumbai, various traditional forms of production including craft activities like pottery, broom and basket making, *jardozi* work, jewellery making, diamond cutting and polishing, etc., are observed to provide a source of earning for a large number of artisans and non-artisans, who have migrated to the city during various phases of city growth. This paper examines the case of the *jari* workers in Mumbai, whose relocation in the competitive metropolitan space gives rise to a complex functioning of market, capital and the manufacturing sector. Based on a questionnaire survey conducted on 75 *jari* workers (75% workers and 25% entrepreneurs) located in the major areas of concentration of *jardozi* in Mumbai, this study, with a historical perspective, analyses the structure of *jardozi* activity and the typology of *jari* workers and focuses on their placement and living conditions as observed in the late 1980s.

The *Jardozi* Craft in Mumbai

Weaving and embroidery works in *jari* and silk thread has been a traditional occupation of many Muslim communities in Mumbai since the early phase of city growth. The Khatri community hailing from Chaul, one of the major ancient trading ports along the Konkan coast, was well known for silk weaving. Silk cloth was used for a variety of brocades, including *kincobs*, *hemrus*, *masurus*, *lapas* and *tas,* which were worked into saris, *cholis*, waistcoats and turbans or *pagdis*. Besides the Muslims, the elite class among the Parsis and the Marathi and Gujarati Hindus like Pathare Prabhus and Bhatias provided a good clientele for these products. The Census of 1901 recorded about 1,000 embroidery and lace workers mainly from the Muslim community in Mumbai. The records of this period also refer to 'jariwala', a distinct class of workers engaged in the recovery of gold and silver from old embroidered clothes, caps, etc. (Govt of Bombay, 1909).

During the pre-industrial phase of evolution of manufacturing in Mumbai, inmigration of various craft communities was promoted by the colonial traders (Edwardes, 1902, Govt of Bombay, 1909). However, during the later phases, many of them had to face subordination and displacement due to the changing policies of colonial trade and administration (Chandavarkar, 1993). Subsequently, *jardozi* work, like many other craft activities in Mumbai, declined during the industrial and urban expansion of the city (Gandhi, 2000). The craft, however, did not become extinct.

In the inter-war period, as manufacturing of *jari* thread was being re-organised, adopting measures of import substitution, trade in the embroidered *jari* products seem to have got polarised into two networks. The first one met the demand of rich Indians and Europeans, while the other was meant for the 'Indian purchasers of humble means' (Gwynne, 1910). The same could be distinguished by size of employment, scale of output, nature of input as well as the designs used by them. The former worked through large-scale production and trading, adapted its designs to suit the tastes of the Europeans and the Anglicised Indian clientele and exercised strict control on the quality of manufacture of *jari*. The latter, operating on a small scale, was more open to the use of imitation *jari* and catered to the needs of a relatively diverse market (Roy, 1999).

As the field information indicates, transformation in *jari* enterprises in Mumbai represented more of the latter type. Due to stagnation and decay of the *jardozi* craft in the Northern cities, the *jari* workers, like many other artisans, migrated out of that region. The city of Mumbai, which by then had emerged as the largest centre of textile mills and other industries and commerce, offered them several advantages. Easy access to cheap imitation *jari* from Surat, few local producers and cheap imported cloth, such as – satin and chiffon, provided a base for survival. The migrant *jari* workers in Mumbai got organised into small workshops to cater to a potential offered by occasion-specific, but more voluminous, demand for relatively cheaper, de-skilled embroidered products. Such a demand not only originated from the lower middle and middle class Gujaratis, but also from the expanding cine industry in the city. Continuous supply of skilled artisans as well as unskilled workers, mainly from Uttar Pradesh and Bihar, but also from

Bengal and Andhra Pradesh, facilitated organisation of the production and trade for a highly competitive market with relatively lower input costs on labour and raw material.

During the post-independence period, the imported varieties of satin or chiffon were increasingly substituted by cheaper local varieties of nylon, terene and synthetic silk. Besides imitation *jari*, use of plastic *jari* as embroidery thread became popular. Similarly, use of artificial pearls, glazed plastic and glass beads, produced locally as well as imported from Australia and Japan, provided a wide range of inputs which facilitated diversification in designs and final output, appealing to the taste of customers from different socioeconomic classes. Opening up of the export markets in Europe, North America and the Middle-East, especially after the 1970s, helped the revival of demand for more skilled embroidery workers. These changes resulted in the mushrooming of small *jari* workshops, which thrived in the newly developing slums and low-class residential areas in the extended parts of the city, in the East as well as North, which had a concentration of Muslim population.

Spatial Organisation and Structure of *Jari* Enterprises in Mumbai in the 1980s

The economic Census of 1980 enumerated 1,331 enterprises employing 6,694 workers engaged in embroidery, by hand and machine along with activities of lace making and carpet and rug weaving. Though insignificant in terms of proportion of industrial workers, these units claimed 8.7 per cent of the total units in the textile industry and recorded an average size of five workers per unit. Unlike some craft activities like pottery, these enterprises primarily absorbed hired workers, who accounted for 70 per cent total workforce. Due to the opening up of the Gulf market for exports, the industry had experienced a boom and its impact could be observed in the late 1980s as well. An estimate based on field investigation indicates that more than 4,000 workers were employed in *jardozi* enterprises in the late 1980s, located in the small clusters in the low-class residential localities of Muslims in Mumbai.

The initial concentration of *jardozi* units was related to factors like proximity to major retail and wholesale markets located in the South-central parts of Mumbai city, including areas like Bhuleshwar, Chakala, Mandvi, Byculla, Nagpada and Kamathipura. The subsequent growth of these enterprises occurred in slums located in the peripheral parts of the main city and their extensions. Thus, the slums of Chitta Camp at Trombay, Shivajinagar and Bainganwadi and other parts of Govandi and Chembur in the East, in Ghatkopar, Vikhroli and Bhandup in the Northeast and Bandra, Jogeshwari, Goregaon and Malad (East) in the West and Northwest have attracted a large number of machine embroidery and hand embroidery units (Figure 1). Findings based on a survey conducted in the 1980s suggests that the clusters in Govandi and Chembur together account for more than 80 per cent of the *jardozi* units in Mumbai (I.C.S.W., 1983).

Structure of the Units

Generally, *jardozi* activity in Mumbai is undertaken in tiny, informal workshops run by an entrepreneur hiring an average of five-six workers per unit. These units are linked with retail and/or wholesale traders who generally operate in the formal sector. The link can be direct with multiple channels or indirect through middlemen with a single channel. Besides regular *jari* embroidery jobs, petty stitching, finishing and peripheral embroidery jobs are also attended to by either the entrepreneurs or the middlemen with the help of home-based tailors and female workers (Figure 2).

Amidst varying demands, trading has become a vital element of control, which the entrepreneurs do not own any more. The main function of the traders is to coordinate the demand for *jari* embroidery work from a variety of sources and articulate the various demands to the entrepreneurs, along with the designs or patterns and the base material. Labour cost

Figure 1

Greater Bombay: Major Areas of Embroidery, Jari Works and Lace Making: Mainly Informal Type, 1980

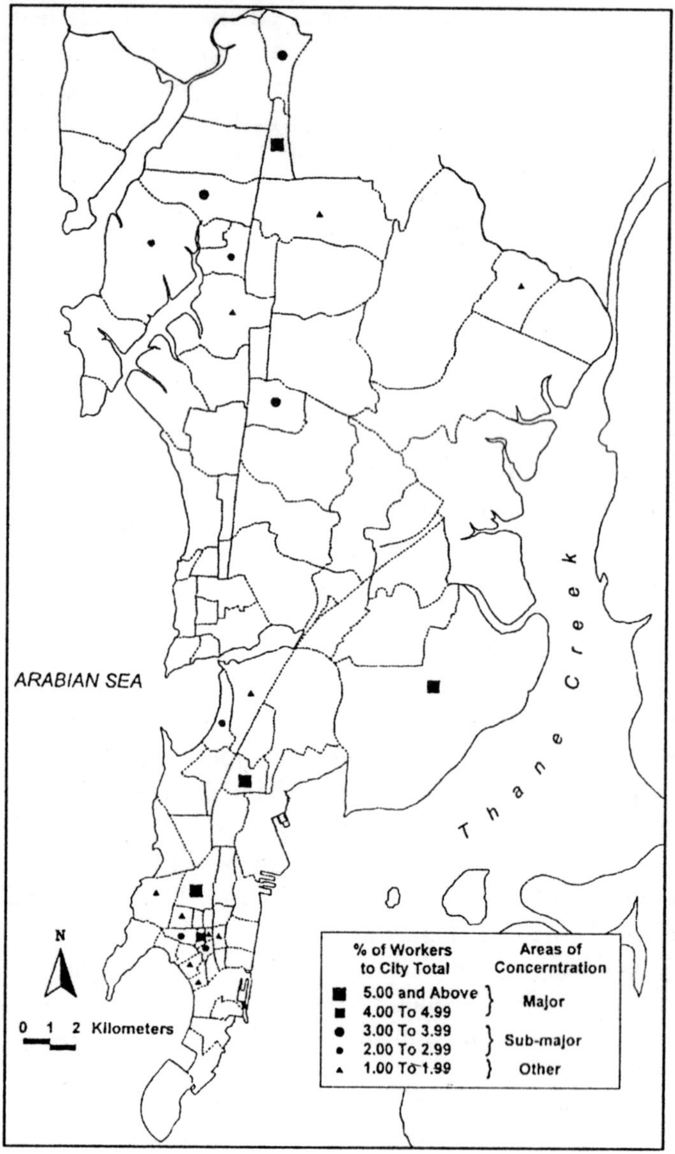

Figure 2
Organisation of Jari Enterprises: Structure, Typology and Linkages

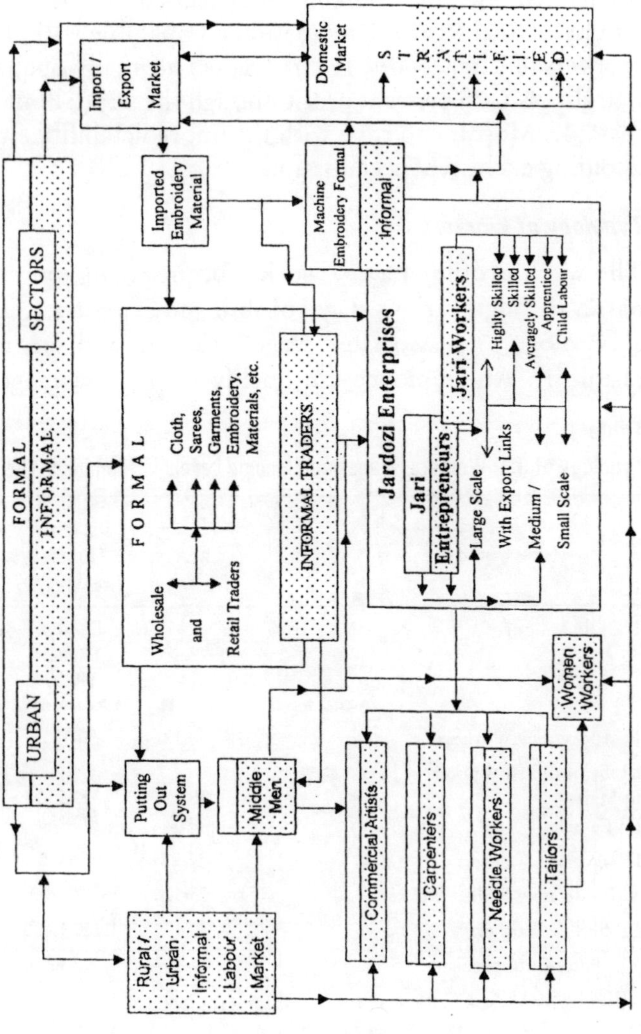

estimates are also negotiated by the traders. By the late 1980s, the role of middlemen became very crucial in the structure that intensified the outsourcing and subcontracting networks among *jardozi* workers, a situation comparable with the artisan workers in Lima (Scott, 1979). Traders normally operate with a margin of 20-25 per cent profit, though the upper limit is unpredictable. Most try to cope with the unpredictability element by avoiding exclusive specialisation.

Typology of Workers

The work profile of a *jari* worker begins as an apprentice and passes through various stages of skill and experience, indicating a typology and associated wage structure and income levels (Table 1). An employee can qualify to be an average worker

Table 1

Typology of Jari Workers, Wages and Income Levels in Mumbai in the 1980s

Typology of Workers/ Entrepreneurs	*Daily Wages in Rupees*	*Average Monthly Income (Rs)*	*% in the Sample*
I. Child Labour	-	Food + Shelter (40-50 Occasional)	5.4
II. Apprentice Workers	10-12	50-300	14.3
III. Semi-skilled Workers			
(i) Males	15-20	300-500	12.5
(ii) Females	8-12	200-250	7.1
IV. Average Workers	18-22	500-800	35.7
V. Moderately Skilled Workers	25-30	800-1,200	6.1
VI. Skilled Workers	30-45	1,100-1,400	3.1
VII. Petty Entrepreneurs	-	1,400-1,500	1.5
VIII. Highly Skilled Workers (Export Link)	50-60	1,400-1,700	4.0
IX. Entrepreneurs – Medium Scale of Output	-	1,700-2,000	4.9
X. Entrepreneurs – Large Scale of Output (Export Link)	-	2,500-5,000	5.4
Total			100

after an apprenticeship of one or two years. However, perfection, excellence and mastery over different types of intricate embroidery work may be acquired over a period of five-six years subject to the skills and overall efficiency of the entrepreneurs with whom they work. The willingness and ability of the entrepreneurs to train an apprentice, especially their access to the market to get high valued, skilled embroidery jobs on a regular basis, are very crucial factors. However, the highly competitive nature of the market and uncertain demand conditions tend to make apprenticeship an exploitative level (Peil, 1979; Allen, 1982), rather than one creating scope for acquiring skills (Hunter, 1973). In a favourable situation, a *jari* worker may climb up from the stage of non-paid child labour without getting any salary to that of highly skilled labour. Only rarely does one become an entrepreneur from this level. Generally, a majority of the *jari* workers stagnate at the level of average skilled worker or petty entrepreneur (Figure 2, Table 1). The typology of the *jari* workers as observed during the late 1980s in Mumbai indicated the hapless condition of young children, who are recruited in large number by entrepreneurs from Bihar, and get a wage of Rs 5-10 per week. The practice reinforces the plight of surplus labour and their depressed wage level. Female workers are generally drawn from the Muslim community and normally allotted the peripheral jobs of either hand stitching or hand embroidery work. Earning an average wage of Rs 8 against labour of four-six hours per day put in as and when the schedule of domestic chores allows, female workers constitute one of the lowest strata in the typology, placed much below the tailors, who earn between Rs 15-30 per day, and only slightly above the child labour. Gender discrimination in the allocation of jobs is quite evident. Empirical evidence from India (Solanki and Qureshi, 1998) as well as other Third World countries (Kowarick, 1979; Scott, 1979) indicate that increasing commercialisation of craft activities not only lowers the wage level of the female workers, but also eases them out. The case of female bidi workers in Mumbai corroborates these findings (Gandhi, 2000).

The above typology indicates that the wage structure and earnings of *jari* workers are affected by the aspects of skill, gender and age, on the one hand, and market and demand conditions, on the other. With one unpaid holiday per week and no paid holidays at any time of the year, *jari* workers normally work for an average of 22-24 days in a month, reduced to 15-18 days during the lean period. During the peak period, the workers may work in two shifts for 10-16 hours per day. An average skilled worker thus normally earns a monthly income of Rs. 500, which, only on a few occasions, increases to more than Rs. 800 on account of double shifts.

The Economics of *Jari* Enterprises

A labour intensive, petty manufacturing activity, *jardozi* work involves negligible capital outlay on equipment and other infrastructure, together amounting to about 11-12 per cent of the total initial investment. About three-four elongated wooden frames of different sizes needed to stretch the embroidered material and a couple of tubelights and ceiling fans are enough to start a *jari* workshop. The most significant of all the items is the operational space, which accounts for more than 80 per cent of the total initial investment. The capacity to either own or rent at least one elongated room with an area of 12 m^2 is one of the crucial factors that distinguishes a worker from an entrepreneur. Except for a small percentage of workshops (15-16%), located in dingy *chawls* in the densely populated lower class areas, and a few spacious ones (4-5%), located in the elite residential areas in South Mumbai, the majority of the workshops (>80%) are located in slums. The latter are mostly iron and wooden shed structures with low roofs, simultaneously serving as residences for the workers and, at times, for the entrepreneurs.

The running expenditure of a *jari* workshop includes 3-5 per cent of the total, spent on rent, transport and electricity.

Labour cost accounts for a lion's share, i.e. 70-80 per cent of the total. Only about 7-8 per cent of the sale proceeds of the entrepreneur go to make the profit. Actually, it is the trader who enjoys an advantage in the total transaction and earns considerably high profits, leaving the risk and uncertainty to be borne by the workers and entrepreneurs.

Socioeconomic Status of *Jari* Workers

Jardozi workers are inter-state migrants, primarily hailing from North India. In the present sample, about 87 per cent of the entrepreneurs and 75 per cent of the workers are migrants. Together, they account for 81 per cent of the total respondents, among which 66 per cent migrated in the 1980s and 11 per cent, before the 1970s. From among the pre-1970 migrants the majority are entrepreneurs with a longer history of work. As mentioned before, the Northern states of Uttar Pradesh and Bihar are the predominant source areas, accounting for 35 per cent and 31 per cent of the total migrant *jari* workers, respectively. West Bengal and Tamil Nadu each account for only 11 per cent of the migrant workers.

Almost all the migrant workers were victims of the 'push' factors of migration. About 25 per cent were displaced from the formal sector and employment in the retail trade or personal services in small towns and had no other choice but to migrate to Mumbai to join the informal sector. Their gross earnings at their native place were, however, as low as Rs 100-250 per month. Not a single rural migrant came from the formal sector.

With an average of six to seven members, the household size of the *jari* workers is moderately large. The average age of both *jari* workers and entrepreneurs is found to be as low as 20. The upper limit of the age of the entrepreneurs is 45. Long working hours spent in poorly lit workshops in awkward sitting postures generally invite numerous health problems, such as weakening of eyesight or acute backache even at a young age. Consequently, many workers become incapable of undertaking skilled jobs even before they reach their forties.

Conditions of poor diet and malnutrition aggravate their illness, leading to premature dismissal from work. Preponderance of child labour, low educational status and perpetuating poverty are the characteristics of the *jari* workers. More than 60 per cent of the *jari* workers are either illiterate or only barely literate. The condition of the entrepreneurs is only marginally better. About 60 per cent of the entrepreneurs are found to have undergone formal education for four to ten years.

During the late 1980s, the average per capita earnings of a *jardozi* entrepreneur were Rs 375, while that of a worker was as low as Rs 205. About 30 per cent of the households recorded having a secondary source of earning besides *jari* work. The monthly expenditure of an average *jari* worker came to Rs 575. The range, however, went as high as Rs 1,669. About 47 per cent of the expenses were on food and clothing and 20 per cent generally sent as remittances.

The plight of the *jari* workers is extremely varied. While young and female workers were found to be the most exploited lot, apprentice and moderately skilled workers were hardly better. Except for the highly skilled workers and entrepreneurs operating on a larger scale, the rest of the *jari* workers seemed to live in conditions of perpetuating poverty and indebtedness. In a family with one earner and 4-5 dependent members, staying either in Mumbai or at the native place, the meagre income of Rs 1,000 to Rs 1,200 appeared to be extremely inadequate. Considering the seasonality and fluctuations in the market and their vulnerability vis-à-vis the traders, entrepreneurs and skilled workers were found to experience considerable income uncertainties. A *jardozi* worker in Mumbai in the 1980s was thus placed well below the poverty line as defined by the prevailing norms (Banerjee-Guha, 1989).

Placement of *Jari* Workers in the Traditional Craft Scenario

A comparative analysis of the artisan communities in Mumbai, including potters, weavers, bidi rollers, basket and broom makers, *jari* workers, jewellery and diamond workers, etc., shows variability in their relative placement. Under the precariously competitive condition of the bipolar economy of the metropolis, they present differential responses and degrees of adaptation to the forces of city growth and changing conditions of demand and technology. The early entrants among the potters from Saurashtra, for instance, could transform themselves as wage earners or petty producers in the metal workshops in the South Central parts of the city. Some, however, were displaced and took to brick making, located in adjoining areas such as Thane and Kalyan on the mainland. Their contemporaries, the Padmashali weavers from Andhra Pradesh and other parts of the Deccan, were initially absorbed as casual workers in textile mills. Some took to bidi rolling till it sufficed for survival and subsequently joined small handloom and powerloom units including dyeing and printing workshops, located in the textile mill area of the city. Some of the later entrants like the potters and broom-makers in Dharavi and the basket-makers in Govandi were affected by the conditions of geographical inertia on the one hand, and social and economic stagnation on the other. They continue with their traditional products as long as there exists a demand for their products from the poorer sections (Gandhi, 2000).

Jari workers have found ways of adapting to the changing market conditions and gradually introduced elements of diversification and mass production. Unlike potters, they do not face any danger, either of losing control on the sources of input or drastic reduction of demand from the local markets (Ibid., 2000). Increase of child and female labour, the subcontracting system and the increasing use of a wide range of inputs to suit the requirements of the segmented market at home and abroad, have provided the opportunity for a certain amount of growth, though not without a risk of fluctuations in demand. It needs, however, to be mentioned that the autonomy of the *jardozi* workers vis-a-vis traders has been considerably subdued over the years. Among the traditional artisans, only the jewellery workers have been able to get the real benefit of an expanding market and establish a comparative stability in their products. The restrictions faced by community specificity and fashion elasticity are not applicable to jewellery workers. Similarly, possibilities of introducing mass production techniques and, thereby, de-skilling the output is also difficult in jewellery work, which again safeguards the status of these workers in the labour market (Gandhi, 2000). Despite having a relatively better adapting capacity, a large number of *jari* workers, in contrast, is found to remain in the occupation, however modest their earnings. Inflow of migrants from Muslim families located in the poorer areas of Uttar Pradesh and Bihar continues to contribute towards a labour surplus condition, leading to a perpetual lowering of real wages and increasing fragmentation of the work. *Jari* workers, unlike *jari* traders, hence remain a socially and spatially marginalised group in the globalising metropolitan environment of Mumbai.

References

Allen, R. (1982): 'Capitalist Development and the Educational Role in Nigerian Apprenticeship', *Comparative Education*, 18(2), 123-137, as in F. Fluitman (ed.) (1989): *Training for Work in the Informal Sector*, ILO, Geneva.

Banerjee-Guha, S. (1989): 'Reachability of the Sites and Services Projects to the Targets Groups in Greater Bombay', *Nagarlok*, XXII (2), 44-57.

Chandavarkar, R. (1993): *Origins of Industrial Capitalism in India – Business Strategies and the Working Classes in Bombay, 1900-1940*, Cambridge University Press.

Edwardes, S.M. (1902): *The Rise of Bombay: A Retrospect*, Times of India Press, Mumbai.

Gandhi, S. (2000): 'Spatio-sectoral Patterns of Informal Manufacturing in Greater Bombay in the 1980s', Unpublished Ph.D. Thesis, Department of Geography, University of Mumbai.

Govt of Bombay (1909): *Gazetteer of Bombay City and Island*, Vol. I, Time Press, Mumbai (Fascimile Reproduction by the Gazetteer Department, Govt of Maharashtra, Mumbai, 1978).

Gwynne, C.W. (1910): Monograph on the Manufacture of Wire and Tinsel in the United Provinces, Allahabad, as in T. Roy (1999) *op. cit.*, 122.

Hunter, G. (1973): 'A Comment on Educational Reform and Employment in Africa', in P. Foster and J.R. Sheffield (eds), *The World Yearbook of Education and Rural Development*, Evans Brothers Ltd., London, 51-58.

International Council on Social Welfare (1983): *The Urban Dead-End? Pattern of Employment in Slum Dwellers*, Somaiya Pub., Mumbai.

King, K. (1979): 'Petty Production in Nairobi: The Social Context of Skill Acquisition and Occupational Differentiation', in R. Bromley and C. Gerry (eds), *Casual Work and Poverty in Third World Cities*, John Wiley and Sons, New York, 217-28.

Kowarick, L. (1979): 'Capitalism and Urban Marginality in Brazil', in R. Bromley and C. Gerry(eds), op. cit., 69-86.

McGee, T.G. (1979): 'The Poverty Syndrome: Making Out in the Southeast Asian City', in R. Bromley and C. Gerry (eds) op. cit., 45-68.

Peil, M. (1979): 'West African Urban Craftsmen', *Journal of Developing Areas*, 14(1), 3-22.

Roy, T. (1999): *Traditional Industry in the Economy of Colonial India*, Cambridge University Press.

Scott, A.M. (1979): 'Who are the Self-Employed', in R. Bromley and C. Gerry (eds), op. cit., 105-32.

Solanki, S.S., and Qureshi, M.A. (1998): 'Village Pottery Industry: A Case Study', Kurukshetra, XLVI(5), 3-15.

18

Urbanisation and Related Issues in the Mumbai Metropolitan Region, 1961-2001

V.S. Phadke

Introduction

Urbanisation has always created new economic opportunities and, hence, spatial concentration of employment avenues, population and socioeconomic infrastructure; this has led to depopulation of the distant countryside (Linsky, 1965). The problem becomes severe in the case of cities experiencing uncontrolled growth and finally attaining metropolitan status; it very often reaches a crisis situation as the infrastructure cannot keep pace with the demand (Mountjoy, 1968). In metropolitan cities with a colonial background, the problem has often been accentuated as external forces have primarily driven their attainment of a metropolitan status. Local entrepreneurs have often been roped in by the colonial masters to facilitate the process of exploitation of cheap labour that needed employment during the agricultural off-season (Ginsburg, 1973). Spatial expansion in function has been facilitated by mass transit system; rise in land values, crowding and congestion in the city itself have made their own contribution (Berry and Kasarda, 1977). This has given rise to rich residential areas of the 'suburbs' mostly marked by the young, dynamic element (Johnson, 1974).

Growth of the metropolitan city was selective and primarily centripetal in character during the colonial period. After independence, too, the city has been found to be the only viable place with a suitable, ready to use infrastructure for further economic expansion. It has thus continued to attract capital because of renewed euphoria and confidence (McGee, 1971). This has led to unbridled growth as, even in the planners' eyes, growth was important above all else. Suburbs started expanding fast and sent a spillover into even the surrounding rural areas on a selective basis, producing the corridor effect (Bogue, 1950). The above process has also segmented the metropolitan community 'spatially' (Pryor, 1968). The planning authorities have contributed their own mite, the primary aim being to tackle immediate problem by developing infrastructure in areas 'adjacent' to metropolitan cities and, thereby, creating resources for future development (Whitehand, 1967). Creation of development inputs in these areas has given rise to the new urbanisation wave.

Today, metropolitan areas are passing through yet another phase of growth inducement, pushed primarily by globalisation forces. Since cities are the organising nuclei, they are trying to become global and vying to develop facilities and infrastructure that would make them attractive to global capital; their main aim is to forge a link with the other countries and cities of the world. In the process, they are also spreading their tentacles into the adjoining rural areas, which are fast getting urbanised (Ginsburg et al., 1991).

Mumbai, a colonial metropolitan city, has also passed through the abovementioned stages. Today, one finds that it has reached saturation point, which is forcing it to expand its activities into the surrounding areas. Initially, this spillover population and function sought locations in the existing urban centres, but some new towns also emerged. Of late, efforts have been made to plan new urban areas or incorporate vast rural areas into the urban fold. The process of urbanisation in the metropolitan region is thus interesting. The focus of this paper is on understanding this process and the related issues that need serious attention.

Evolution of the Mumbai Metropolitan Region (MMR)

Situated on the Western seaboard of India and backed by the Sahyadrian wall to its East, Mumbai should have easily developed connections to the North and the South rather than to the East. Being insular in character, however, the connection to the South was constrained by its physical separation from the mainland by a wide creek. Moreover, during the colonial times, better links were forged with the East because of the rich and extensive hinterland in that direction; this was achieved by taking advantage of the gentle slopes along the river valleys and passes carved out in the Sahyadrian wall by headward erosion. It is only later that more lasting contacts were established with the North and only recently with the South.

Deep waters on the protected side of the islands and connections with the hinterland made Mumbai an assembly point initially and later a processing centre (Nairne, 1894). Capital and labour soon moved into it and the settlement started growing, albeit slowly at first, as a lot of reclamation was needed. This laid the foundation of its Northward expansion. Railways and roads provided the mass transit system. The growth was initially restricted to some extent by the Mahim and Sion creeks, but the growing streams of migrants had to find refuge in Salsette, lying immediately to the North, transforming the rural settlements around the local railway stations into suburbs (Deshpande, 1974; Sita and Phadke, 1984).

Continuous growth of the city, even after independence, made the suburban boundary look too restrictive (Brahme, 1978); unhindered growth of functions and, thereby, population made it essential for the city to encroach upon the surrounding rural areas. The process was, however, selective, encouraged by route corridors running parallel to river valleys or the shoreline and discouraged by unfavourable relief, with the resulting lack of access (Phadke and Sita, 1981a). Penetration along the route sectors depended on historical ties and status of the settlements connected, frequency of the transport services provided, age of the route, etc. (Phadke and Sita, 1981b). The emergence of new urban centres was thus rather spatially uneven. The city occasionally pushed its functions on to the distant periphery, sometimes voluntarily, but on many occasions by force (Koundinya et al., 1967). The process was intended to relieve pressure on the former and help generate growth in the latter.

The liberalisation policy of the 1990s has infused new blood into planning and the city is now trying to catch up with the global trends of becoming a service centre (BMRDA, 1995). Naturally, new infrastructure for faster transport and communication is being added. Highways are being modernised to make smooth arteries by concretising and constructing flyovers to avoid delays at traffic signals. Telephone services are being multiplied and Internet services extended; extension programmes for awareness and literacy are also being planned.

Data Base and Method of Analysis

The analysis in the paper is based on the data published by the Census. Figures related to population of towns, falling within the jurisdiction of the Mumbai Metropolitan Region (MMR), workers belonging to different non-agricultural categories enumerated in these towns, etc., were used to derive certain simple statistics, such as growth rate of population of urban centres or percentage share of each corridor in total urban population. This has enabled appreciation of the pattern of growth within and outside Greater Mumbai. The spatio-temporal pattern of urbanisation was analysed with the help of maps showing the population of the towns during the different Census years. The status of the various towns in each Census year was also noted and cross-tabulated with their growth rate in the following decade; this has helped in understanding the relationship between the two. The change in the population of the towns during different periods has been correlated to that in workers belonging to different non-agricultural categories during the corresponding periods. The Census of 1981 gives only truncated occupational data, while that of 2001 has not so far released any figures related to workers. The correlation exercise was, therefore, limited to only two periods, viz. 1961-71 and 1971-91.

Results of Analysis

The **urban population** of the MMR, inclusive of the city of Greater Mumbai, has increased from 46.62 lakhs in 1961 to 179.68 lakhs in 2001 (Table 1 A). The City[1] continues to enjoy primacy in terms of urban population, although the rest of the region is slowly picking up in this respect. The fact that the City population has not even trebled during the period of 40 years, but that of the rest of the region has increased more than tenfold, is testimony to this phenomenon. While the absolute growth in population of the City is steady, hovering at around 20 lakhs per decade, the addition to the urban population of the region outside Greater Mumbai shows rapid strides only during the last two decades (Table 1 B); 80 per cent of the total growth in urban population of the latter has taken place during this period. The City, on the other hand, has shown signs of saturation. The growth rate of the population in the City appears to be only modest during the first two decades and just equivalent to the average population growth for the country as a whole during the rest of the period (Table 1 C). The growth rate of the urban population in the rest of the region has always outstripped that of the City and the decade, 1981-91, appears to be the period of the most rapid growth there; this needs careful interpretation, as it is primarily the result of the low population base, especially in the initial stages. The region outside the City contributed only modestly to the growth of urban population during the first two decades, but its share is more than matching during the last 20 years (Table 1 D). However, even this arises out of the saturation in the City and the resulting spillover.

Table 1

(A) Mumbai Metropolitan Region – Urban Population in Different Segments, 1961-2001

Year/Segment	*1961*	*1971*	*1981*	*1991*	*2001*
MMR	46,62,182	68,33,732	98,17,464	1,34,53,646	1,79,67,673
GM	41,52,056	59,70,575	82,43,405	99,25,891	1,19,14,396
MMR-GM	5,10,126	8,63,157	15,74,059	35,27,755	60,53,277

(B) Mumbai Metropolitan Region – Urban Population Growth in Different Segments, 1961-2001

Period/Segment	*1961-71*	*1971-81*	*1981-91*	*1991-2001*	*1961-2001*
MMR	21,71,550	29,83,732	36,36,182	45,14,027	1,33,05,491
GM	18,18,519	22,72,830	16,82,486	19,88,505	77,62,340
MMR-GM	3,53,031	7,10,902	19,53,696	25,25,522	55,43,151

(C) Mumbai Metropolitan Region – Growth Rate of Urban Population in Different Segments, 1961-2001

Period/Segment	*1961-71*	*1971-81*	*1981-91*	*1991-2001*	*1961-2001*
MMR	46.58	43.66	37.04	33.55	285.39
GM	43.80	38.07	20.41	20.03	186.95
MMR-GM	69.20	82.36	124.12	79.55	1,086.62

(D) Growth Share of Mumbai Metropolitan Region outside Greater Mumbai, 1961-2001

1961-71	*1971-81*	*1981-91*	*1991-2001*	*1961-2001*
16.26	23.83	53.73	55.95	41.66

The spatial pattern of urbanisation in the metropolitan region is not uniform (Figure 1 A-E). While the differentiation between the route and non-route sectors is to be expected, besides the distance decay, one also finds a temporal change in the pattern along different routes. The Mumbai-Pune railway was the only sector promoting urbanisation in the initial stages (Figure 1 A); the Mumbai-Pune/Konkan and Mumbai-Agra roads played second fiddle; this is partly due to the nature of the route and in part to the 'age' (Phadke and Sita, 1981b). Even

Figure 1
Mumbai Metropolitan Region: Urban Population

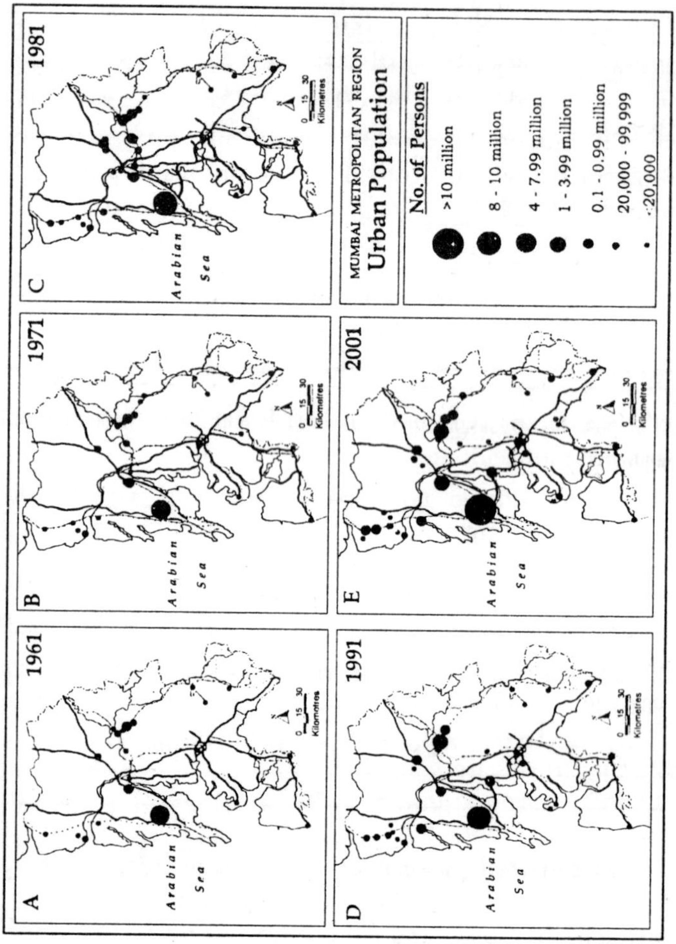

Table 1

(A) Mumbai Metropolitan Region – Urban Population in Different Segments, 1961-2001

Year/Segment	*1961*	*1971*	*1981*	*1991*	*2001*
MMR	46,62,182	68,33,732	98,17,464	1,34,53,646	1,79,67,673
GM	41,52,056	59,70,575	82,43,405	99,25,891	1,19,14,396
MMR-GM	5,10,126	8,63,157	15,74,059	35,27,755	60,53,277

(B) Mumbai Metropolitan Region – Urban Population Growth in Different Segments, 1961-2001

Period/Segment	*1961-71*	*1971-81*	*1981-91*	*1991-2001*	*1961-2001*
MMR	21,71,550	29,83,732	36,36,182	45,14,027	1,33,05,491
GM	18,18,519	22,72,830	16,82,486	19,88,505	77,62,340
MMR-GM	3,53,031	7,10,902	19,53,696	25,25,522	55,43,151

(C) Mumbai Metropolitan Region – Growth Rate of Urban Population in Different Segments, 1961-2001

Period/Segment	*1961-71*	*1971-81*	*1981-91*	*1991-2001*	*1961-2001*
MMR	46.58	43.66	37.04	33.55	285.39
GM	43.80	38.07	20.41	20.03	186.95
MMR-GM	69.20	82.36	124.12	79.55	1,086.62

(D) Growth Share of Mumbai Metropolitan Region outside Greater Mumbai, 1961-2001

1961-71	*1971-81*	*1981-91*	*1991-2001*	*1961-2001*
16.26	23.83	53.73	55.95	41.66

The spatial pattern of urbanisation in the metropolitan region is not uniform (Figure 1 A-E). While the differentiation between the route and non-route sectors is to be expected, besides the distance decay, one also finds a temporal change in the pattern along different routes. The Mumbai-Pune railway was the only sector promoting urbanisation in the initial stages (Figure 1 A); the Mumbai-Pune/Konkan and Mumbai-Agra roads played second fiddle; this is partly due to the nature of the route and in part to the 'age' (Phadke and Sita, 1981b). Even

Figure 1
Mumbai Metropolitan Region: Urban Population

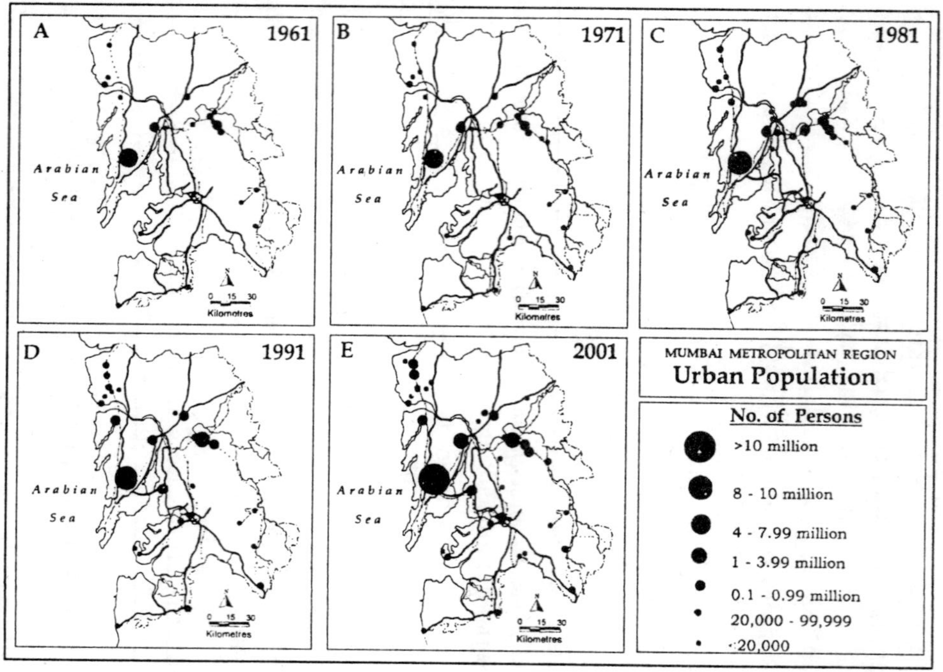

here, one finds distance decay, with accessibility to the City becoming an important consideration. The distance, however, is not simply geographical and other considerations, such as frequency of train services or availability of urban infrastructure, have played an important role. Other sectors show only a nominal presence of urban population, focusing on some old urban nodes like Vasai, Alibag or Uran. The first corridor outside the traditional sectors to show some semblance of growth was the Western Railway route, where the nodes around railway stations emerged and began to come into prominence (Figure 1 B, C); even here, the growth was only modest. Surprisingly, the third corridor to the East, which was opened to traffic in the early 1970s, did not really pick up even after nearly a decade in 1981. This was primarily due to the faulty procedure of designation of settlements as 'towns' (Rahinj and Phadke, 1982); only Ghansoli was identified as a town by the Census in 1981 and Vashi, despite being the first residential node of Navi Mumbai, was not considered a town. Panvel, too, showed only modest growth during this period of the initial teething troubles. The third corridor as well as the Western Railway sector really picked up during the last two decades (Figure 1 D, E); it is primarily attributable to the granting of urban status to Navi Mumbai in Thane and Raigad districts, improved access and saturation of the traditional route sectors. One also observes the process of amalgamation of urban areas by merging some of the small towns and even the adjoining rural areas into the nearby parent city to form municipal corporations; four such corporations have been constituted in 1991 (Department of Geography, 2003). The share of the various sectors in urban population, during the various Census years, gives a quantitative dimension to these spatio-temporal trends (Table 2).

The metropolitan region shows ups and downs in the number of towns (Table 3). This is due to policy factors such as the decision to designate some of the rural settlements as towns or to amalgamate small, old towns, along with the adjoining rural areas, into municipal corporations. The former was very

Table 2
Percentagewise Share of Different Route Sectors in Urban Population of Mumbai Metropolitan Region outside Greater Mumbai, 1961-2001

Sector/Year	*1961*	*1971*	*1981*	*1991*	*2001*
Mumbai-Pune Railway	72.54	70.69	69.73	62.67	54.10
Mumbai-Agra Road	9.34	11.78	15.20	11.38	11.17
Mumbai-Pune/Konkan Road (Old)	9.37	9.62			
Western Railway	8.75	7.90	6.71	11.09	17.16
Mumbai-Pune/Konkan Road (New)			8.36	14.85	17.57
Mumbai-Ahmedabad Road	-	-	-	-	-
Total	100	100	100	100	100

Table 3
Mumbai Metropolitan Region – Distribution of Towns Based on Status, 1961-2001

Year/Status	*1961*	*1971*	*1981*	*1991*	*2001*
Mega City					1
Other M. Cities	1	1	1	2	2
Other Class I	2	2	5	5	9
Class II	1	4	1	3	3
Class III	3	3	11	5	5
Class IV	3	8	8	5	9
Class V	10	6	9	6	8
Class VI	1	3	3	1	-
Merged				16	2
Declassified			1	1	1
Total	21	27	38	27	37

prominent in 1971 and 1981 and the latter, in 1991. This can be seen from the increase in the number of towns from 21 in 1961 to 38 in 1981. In 1991, on the other hand, as many as 16 small towns were merged into nearby parent cities; four municipal corporations, viz. Thane, Mira-Bhayandar, Bhiwandi and Kalyan, have been formed this way. In the same year, two new towns, viz. Navi Mumbai-Thane and Raigad segments – have been constituted, taking the tally of Class I and II towns up. The process is extended into 2001 and one new municipal corporation, viz. Navghar-Manikpur, was constituted; several small rural settlements such as Waliv and Kophrad along the Western Railway line, Katai and Padgha along the Agra road, and Ambivali and Mohapada off the Mumbai-Pune/Konkan road have also been elevated to urban status due to their residential/industrial significance.

Simultaneously, one also finds a trend towards the breaking up of the existing municipal corporations; Ambarnath and Badlapur, for example, have been carved out as separate urban areas from Kalyan Municipal Corporation; the rest of Kalyan has been given the identity of Kalyan-Dombivli Municipal Corporation. The number of towns in the higher classes has thus gone up, while that of those in the lower classes has not necessarily gone down.

The growth of towns belonging to different classes is not uniform (Table 4). Generally, the medium sized towns have shown a faster growth compared to large and small towns during 1961-71. The lower rate of growth in large towns is due to saturation, while that of small towns is due to their poor resource base. Although the pattern remains the same for the periods 1971-81, 1981-91 and 1991-2001, one finds

Table 4

Mumbai Metropolitan Region – Growth Rate of Towns of Different Status

(A) 1961-71

Class/G.R. (%)	*100+*	*50-100*	*20-49.9*	*0-19.9*	*Total*
I		2	1		3
II			1		1
III		2	1		3
IV	1		2		3
V		4	4	2	10
VI				1	1
Total	1	8	9	3	21

Contd...

Contd...

(B) 1971-81

Class/G.R (%)	*100+*	*50-100*	*20-49.9*	*0-19.9*	*Declassified*	*Total*
I		2	1			3
II	1	1	2			4
III	1		1	1		3
IV	2	3	2	1		8
V	1	1	2	2		6
VI		1		1	1	3
Total	5	8	8	5	1	27

(C) 1981-91

Class/G.R(%)	*100+*	*50-100*	*20-49.9*	*0-19.9*	*Declin-ing*	*Merged*	*Declassi-fied*	*Total*
I			1	1		4		6
II						1		1
III	1	1	1	1		7		11
IV			1	2		5		8
V			1	2	2	4		9
VI			1			1	1	3
Total	1	1	5	6	2	22*	1	38

* This figure does not tally with that in Table 3, as it is not possible to calculate the growth rate of even the parent cities into which the smaller towns have been merged.

(D) 1991-01

Class/ G.R. (%)	*100+*	*50-100*	*20-49.9*	*0-19.9*	*Declin-ing*	*Merged*	*Declassi-fied*	*Total*
I	2	2	2	1				7
II	2	1						3
III			4			1		5
IV		2	1	1		1		5
V	2	1	1		1		1	6
VI				1				1
Total	6	6	8	3	1	2	1	27

exceptionally high growth of medium towns during 1971-81 and that of large towns during 1991-2001. The latter is due to the growth of residential satellites like Mira-Bhayander, Ulhasnagar and Bhiwandi, which were medium sized towns in the preceding decade and became large in the later decade.

Table 5

MMR – Absolute and Relative Change in Population and Number of Workers in Different Non-agricultural Categories, 1961-71 and 1971-91

Population/ Workers	*Absolute 1961-71*	*Absolute 1971-91*	*Relative (%) 1961-71*	*Relative (%) 1971-91*
Population	2171550	6471154	46.50	87.33
Category III	-2829	10737	-9.17	34.81
Category IV	7039	30476	25.99	112.52
Category V	298483	676687	40.90	92.72
Category VI	25564	141249	52.36	289.30
Category VII	207509	549758	62.62	165.91
Category VIII	55871	235524	26.74	112.72
Category IX	26105	474879	5.52	100.50

The categories here are as per the 1961 Census

Increase in urban population has led to significant growth in the number of workers in the different non-agricultural categories, except in the less important occupational category, i.e. mining and quarrying (Table 5) and this is easily explicable. It is noteworthy, however, that the gain is more than proportionate in trade and commerce and manufacturing during 1961-71. During the period, 1971-91, all the categories show gain, but it is meagre in mining and quarrying and phenomenal in construction, suggesting a construction boom; manufacturing is growing only sluggishly and so are the other secondary and tertiary sectors, except perhaps trade and commerce. It is also interesting to note that the inter-correlation values between the change in population of towns and change in number of workers as also those among workers engaged in different categories are high during 1961-71 (Table 6). These values continue to remain high in 1971-91, except for category III, which shows low values. Surprisingly, change in workers in this category shows a positive relationship

Table 6
MMR: Correlation Matrix for Changes in Population and Workers in Different Non-agricultural Categories

(A) 1961-71

	Occupational Categories							
	Pop. Ch.	*III*	*IV*	*V*	*VI*	*VII*	*VIII*	*IX*
Pop. Ch.	1.0000							
III	-0.9002	1.0000						
IV	0.9971	-0.9049	1.0000					
V	0.9993	-0.8968	0.9969	1.0000				
VI	0.9985	-0.9009	0.9949	0.9966	1.0000			
VII	0.9997	-0.9017	0.9958	0.9983	0.9987	1.0000		
VIII	0.9990	-0.9014	0.9961	0.9975	0.9978	0.9989	1.0000	
IX	0.9719	-0.8788	0.9742	0.9730	0.9727	0.9694	0.9690	1.0000

The categories here are as per the 1961 Census

(B) 1971-91

	Occupational Categories							
	Pop.Ch.	*III*	*IV*	*V*	*VI*	*VII*	*VIII*	*IX*
Pop. Ch.	1.0000							
III	0.2015	1.0000						
IV	0.9928	0.1641	1.0000					
V	0.9743	0.2226	0.9539	1.0000				
VI	0.9991	0.2070	0.9914	0.9668	1.0000			
VII	0.9974	0.1873	0.9960	0.9578	0.9973	1.0000		
VIII	0.9987	0.1916	0.9929	0.9634	0.9997	0.9978	1.0000	
IX	0.9983	0.1861	0.9957	0.9604	0.9987	0.9989	0.9993	1.0000

The categories here are as per the 1961 Census

with both change in population and change in number of workers belonging to other non-agricultural categories. The initial decrease in workers belonging to this category is probably in response to the availability of alternative job opportunities in the other sectors, whereas the increase in later years is partly due to the substantial increase in urban population requiring a corresponding increase in construction. The most important activity undergoing a change is thus mining and quarrying, including sand dredging; this is due to increase in construction activity (Phadke and Mukherji, 2003).

An analysis of linkage through the typal structures of McQuitty (Yeates, 1974) reveals that change in population is primarily linked to change in workers in categories related to manufacturing, trade and commerce and transport and communication during 1961-71 (Figure 2 A). This suggests that expansion in these activities is instrumental in urbanisation, these being the main city forming activities. Growth in workers in construction and other services is a natural concomitant. Occupational category III has been isolated because of

Figure 2

Typal Structures Indicating the Relationship between Change in Population and that in Workers Belonging to Different Non-agricultural Categories

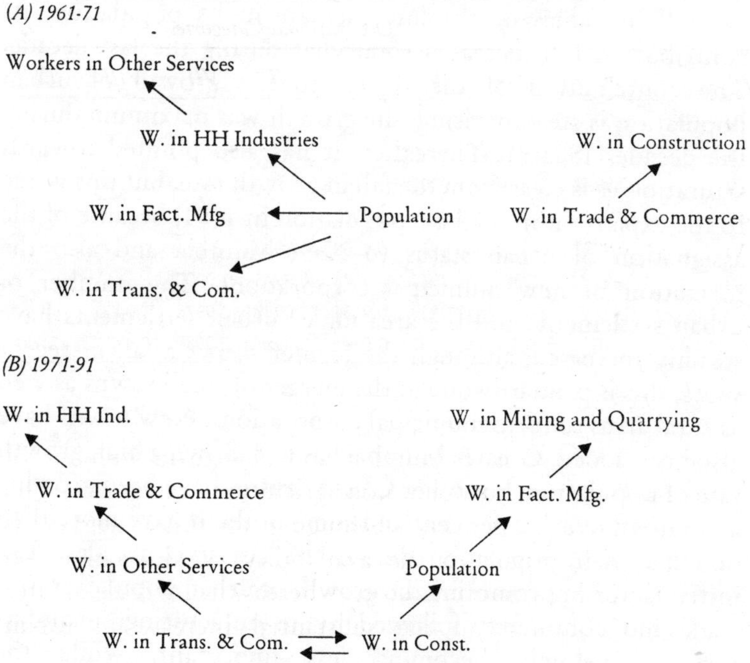

negative values and the initial fall in these activities is related to the switchover to newly emerging, more remunerative activities. During the period, 1971-91, factory manufacturing continues to remain a strong contender in bringing about a change in urbanisation, but other causal sectors, especially construction, transport and communication and other services are nearly equally important in their appeal (Figure 2B). Change in workers in the construction sector has come closer to population change, suggesting the urgency due to the rapidly increasing demand for new roads and buildings; this also explains the positive relationship of change in workers in quarrying and mining to population change.

Main Findings

The foregoing analysis reveals that the MMR outside Greater Mumbai has steadily increased its share in urban population. The city has shown signs of saturation, which is apparent from the fall in values of absolute increase in its population; its contribution has increased somewhat during the last decade. The contribution of the region to the growth of urban population is steadily rising; the growth was maximum during the decade, 1981-91. Thereafter, it has also pointed towards saturation as is clear from the fall in growth rate, but this is due to the expansion in the base population in 1991, because of the assignation of urban status to Navi Mumbai and also the formation of new municipal corporations. The number of urban settlements and the area under urban settlements have steadily increased; although the former shows a fall in recent years, this is primarily due to the merger of small towns as well as rural areas to form municipal corporations. New towns have also been added. Greater Mumbai has not shown a high growth rate of population, but other Class I centres have been growing at a rate of over 50 per cent and some of them have more than doubled their population during the last two decades. The initial factor in promoting the growth of urban population was trade and commerce, followed by manufacturing; manufacturing is slowly becoming less important, while the communication and service sectors are gaining in importance along with trade. Growth in household industry is also appreciable and it primarily reflects on contract jobs offered in place of regular formal employment. The construction sector has experienced a boom, especially during the period, 1971-91.

Major Issues

Trade and commerce and manufacturing have initially supported the growth of urban population in the area, but the former is becoming less labour intensive and the latter is on the decline. Under these conditions, unemployment is becoming a common feature and, along with that, frustration and crime. A large chunk of the urban growth in the region after the 1980s has taken place due to the sprawl outside the city into the hitherto green rural areas; it is not only threatening the rural environment, but also the society there. Fringe management is a major issue at stake (Phadke, 1990) and only a few communities have shown the will to keep their environment green and peaceful (Phadke and Rao, 2003). The newly added urban population is confined to the same old transport corridors, causing problems both for the existing commuters and the newly added ones. There are attempts to accommodate some of the growing population into the city itself by promoting high-rise buildings, but such attempts have their own risks as they would strain the existing infrastructure further. According to a recent survey of the Economic Intelligence Unit, Mumbai is worse than the Senegalese capital, Dakar (EIU, 2004). It is among the least habitable cities in the world. The development authorities are still sticking to their policy of a city oriented development pattern, while the rural areas are getting neglected. Their current emphasis is on making Mumbai a global city. While this is welcome, it cannot be at the cost of those who have contributed their share to its present glory. A city which is dependent on global as well as intra-regional forces for survival, should pay adequate attention to the urban centres in its immediate periphery. Their survival is vital to the prosperity of the city.

Acknowledgement

The author wishes to place on record the help rendered by Shri V.S. Rao and Dr Dipti Mukherji, Department of Geography, in the preparation of maps for this paper.

Note

1. The term 'City' refers to Greater Mumbai throughout the rest of this paper.

References

Berry, B.J.L., and Kasarda, J.D. (1977): *Contemporary Urban Ecology*, Macmillan, New York.

BMRDA (1995): Draft Regional Plan for Bombay Metropolitan Region, 1996-2011, Mumbai.

Bogue, D.J. (1950): *The Structure of the Metropolitan Community*, University of Michigan Press.

Brahme, S. (1978): 'The Role of Bombay in the Economic Development of Maharashtra', in A.D. Noble and A. Dutt (eds), *Indian Urbanisation and Planning*, First Reprint, Tata McGraw Hill, New Delhi, 313-25.

Department of Geography (2003): 'Marginal Towns of Maharashtra, 1961', Project Report, M. A. Part II, Department of Geography, University of Mumbai.

Deshpande, C.D. (1974): 'The Suburbs of Greater Bombay', Occasional Paper No. 2, Centre of Urban Research, Osmania University, Hyderabad.

Economist Intelligence Unit (2004): 'Survey of Hardship in World Cities', London, as in *The Times of India*, 20th February 2004.

Ginsburg, N. (1973): 'From Colonialism to National Development: Geographical Perspectives on Patterns and Policies', *Annals*, AAG, 63, 1-21.

Ginsburg, N., Koppel, B., and McGee, T.G. (eds)(1991): *The Extended Metropolis: Settlement Transition in Asia*, University of Hawaii Press, Honolulu.

Johnson, J.H. (1974): *Suburban Growth*, John Wiley and Sons, London.

Koundinya, U.K., et al. (1967): *Industries in Bombay Metropolitan Region - A Survey*, Town Planning Valuation Department, Regional Plans, Mumbai.

Linsky, A. (1965): 'Some Generalisations Concerning Primate Cities', *Annals*, AAG, 55, 506-13.

McGee, T.G. (1971): *The Urbanisation Process in the Third World*, Bell and Sons, London.

Mountjoy, A.B. (1968): 'Million Cities: Urbanisation and the Developing Countries', *Geography*, 53, 365-74.

Nairne, A. (1894): *History of Konkan*, Govt Central Press, Mumbai.

Phadke, V.S. (1990): 'Bombay Metropolitan Region: Some Issues in Regional Development', *EPW*, 25(3), 175-78.

Phadke, V.S., and Mukherji, D. (2003): 'Urbanisation and Development, 1961-91: Some Observations', *Transactions*, IIG, 25 (1&2), 116-24.

Phadke, V.S., and Rao, V.S. (2003): 'Stagnant Town in the Fringe Area of a Metropolitan City: The Case of Sandor near Mumbai', Paper presented at the Seminar on 'Management of Urban Fringe', organised by IIG, Pune, and communicated for publication in the Proceedings of the Seminar.

Phadke, V.S., and Sita, K. (1981 a): 'Spatial Pattern of Urban Impact in a Metropolitan Region - A Case Study of Bombay', *Annals*, NAGI, 1(1), 18-28.

Phadke, V.S., and Sita, K.(1981 b): 'Differences between Routeways in Promoting Spread of Urban Characteristics - A Study of Bombay Metropolitan Region', *Deccan Geographer*, 18 (2&3), 109-18.

Pryor, R.J. (1968): 'Defining the Rural-Urban Fringe', *Social Forces*, 47, 202-15.

Rahinj, D., and Phadke, V.S. (1982): 'New Towns of India - 1971: An Appraisal', *Indian Journal of Landscape Systems*, 5 (1&2), 99-103.

Sita, K., and Phadke, V.S. (1984): 'Intra-Urban Variations in Population Distribution and Growth in Greater Bombay, 1971-81', *Urban India*, 4(2), 3-12.

Whitehand, J.R.W. (1967): 'Fringe Belts: A Neglected Aspect of Urban Geography, *Transactions*, IBG, 41, 223-33.

Yeates, M. (1974): *An Introduction to Quantitative Analysis in Human Geography*, McGraw Hill, New York.

Contributors

S.P. Das Gupta, Former Director, National Atlas and Theamatic Maping Organisation, Kolkata.
Resi.: 309, Lake Town, Block A, Kolkata – 700 089.

Parbati K. Sircar (Deceased), Formerly Professor Emeritus of Geography, State University of New York, New Paltz, N.Y. 12561.

Satyesh C. Chakraborty, Retired Professor of Urban Management, IIM, Kolkata.
Resi.: Flat 4, Block 8, 96, Kankulia Road, Kolkata 700 029.

R.G. Gidadhubli, Former Director, Centre for Central Eurasian Studies, University of Mumbai, Mumbai.
Resi.: 4, Devashish Apartment, Narkar Marg, Ghatkopar (E), Mumbai 400 077.

S. Banerjee-Guha, Professor of Human Geography, University of Mumbai, Vidyanagari, Mumbai 400 098.

Jayasree De, Professor of Geography, M.S. University of Baroda, Vadodara 390 002.

Dipti Mukherji, Reader in Geography, University of Mumbai, Vidyanagari, Mumbai 400 098.

V.S. Phadke, Retired Professor of Geography, University of Mumbai, Mumbai.
Resi.: 2/17, Dnyanayog Society, Vazira Naka, L.T. Road, Borivali (W), Mumbai 400 091.

Gopal Krishan, Professor of Geography, Panjab University, Chandigarh 160 014.

Surya Kant, Lecturer in Geography, Panjab University, Chandigarh 160 014.

Anjana Desai, Retired Professor of Geography, Gujarat University, Ahmedabad.
Resi.: A 46, Shree Rang Villa, Vastrapur, Ahmedabad 380 015.

S.C. Mukhopadhyay, Professor of Geography, University of Calcutta, 35, Balligunge, Circular Road, Kolkata 700 019.

S. Ebeneezer, Lecturer (Selection Grade), Department of Geography, University of Mumbai, Vidyanagari, Mumbai 400 098.

Daksha C. Barai, Retired Professor of Geography, Bangalore University, Bangalore.
Resi.: 54, South End Apartments, First Cross, L.I.C. Colony, Bangalore 560 011.

K.M. Kulkarni, Professor and Head, Department of Geography, Gujarat University, Navrangpura, Ahmedabad 380 009.

Sudha Srivastava, Reader in Geography, University of Mumbai, Vidyanagari, Mumbai 400 098.

A.S. Jadhav, Professor of Geography, University of Mumbai, Vidyanagri, Mumbai 400 098.

Heinz Nissel, Professor, Department of Geography, University of Vienna, Austria.

Smita Gandhi, Reader in Geography, University of Mumbai, Vidyanagari, Mumbai 400 098.

Index